I0146458

Beekeeping for Beginners

How To Make Modern
Beekeeping A Successful Hobby

Avery Hansen

© Copyright 2020 - All rights reserved.

The content contained within this book may not be reproduced, duplicated or transmitted without direct written permission from the author or the publisher.

Under no circumstances will any blame or legal responsibility be held against the publisher, or author, for any damages, reparation, or monetary loss due to the information contained within this book, either directly or indirectly.

Legal Notice:

This book is copyright protected. It is only for personal use. You cannot amend, distribute, sell, use, quote or paraphrase any part, or the content within this book, without the consent of the author or publisher.

Disclaimer Notice:

Please note the information contained within this document is for educational and entertainment purposes only. All effort has been executed to present accurate, up to date, reliable, complete information. No warranties of any kind are declared or implied. Readers acknowledge that the author is not engaged in the rendering of legal, financial, medical or professional advice. The content within this book has been derived from various sources. Please consult a licensed professional before attempting any techniques outlined in this book.

By reading this document, the reader agrees that under no circumstances is the author responsible for any losses, direct or indirect, that are incurred as a result of the use of the information contained within this document, including, but not limited to, errors, omissions, or inaccuracies.

Table of Contents

Introduction

It is difficult to imagine a world without the presence of bees. Many of us overlook the significant role that these insects play in building our economies and beautifying nature around us. The magic of bees lies in their ability to pollinate plants and crops, causing them to receive a much needed boost of growth, color, and health. If we had to adjust to a world without bees, we would also have to adjust to not being able to eat most of the raw and whole foods which give us plenty of nutrients. Nonetheless, the idea of bees becoming extinct is one that could become our reality if we do not actively seek to conserve these wonderful species.

The fact of the matter is that bees are threatened in the wild. Some of the threats that they are faced with include diseases, pesticides, and the stresses of commercial beekeeping. When a bee is threatened it cannot function in its optimum capacity and this inevitably threatens our environment, too. We need bees to continue to thrive in order for us as human beings to thrive as well. Beekeeping is a selfless act of conserving the species of bees and beautifying our gardens in the process. It is a suitable hobby for those individuals who love gardening, spending time in nature, and using their hands to serve.

A successful beekeeper is one who enjoys the process of growing a colony of bees, noticing every minor development in the life of these fascinating insects. Just as much as a gardener would enjoy watching their vegetable crops grow to become healthy produce, a beekeeper would be rewarded by tending to their beehives, giving their bees an opportunity to grow and carry out their duties in a safe environment. Of course, when we speak about beekeeping, we cannot forget about the sweet gift that our bees generously offer us for our service in taking care of them. Beekeepers who keep honey bees are able to collect the surplus honey produced by the bees and bottle it for use at home or to sell commercially.

The honey produced by the bees gives beekeepers an opportunity to turn this fulfilling hobby into a small business. Not only can bees produce honey, their beeswax is also a great primary product in making wax products such as candles, body butters, and lip balms. Therefore, those who are looking to earn an extra income are encouraged to continue reading this book in order to learn more on how beekeeping can potentially open entrepreneurial doors in your life. Nonetheless, the products from our bees will be compromised if we do not learn how to take care of these insects and provide an environment free from any possible dangers to their existence. This places a significant responsibility on us as beekeepers to study more about bees, their characteristics, and the necessary conditions that they require in order to become a fully functioning hive.

This book is intended to guide new beekeepers in purchasing their first beehive and nurturing the development of their bees until they can successfully buy another beehive and continue to expand their operations. This book will show you that beekeeping does not require much intervention from the beekeeper; tending to the hive every so often will be enough to ensure that the bees are safe and prospering. However, there is a significant amount of investment that new beekeepers will have to make at the beginning of their journey as they bring the bees home and purchase some of the equipment and accessories required. Even so, this investment will ensure the survival of the bees in the colonies and that they will, with time, increase in number and multiply the potential outputs of honey and other products.

From One Beekeeper to Another

My journey into beekeeping started over 5 years ago as an exciting hobby which offered me an opportunity to make a positive contribution to our environment. Being a lover of nature and the outdoors, I was eager to learn more about this hobby and become hands-on. Learning how to take care of bees was not an overnight process. This is because bees have a life of their own and there are so many dynamics to understand about the beehive and how it operates. I studied a lot of great books and listened to many successful beekeepers in my hopes to acquire as much knowledge as I could.

However, the vast amounts of knowledge which I had acquired only prepared me to a certain degree for what I would experience upon starting my journey. I found that there were so many elements to beekeeping which I was never warned about as an amateur beekeeper. A great deal of what I know.now and what I will generously share with you in this book, was learned on the job through sheer experience. It took making mistakes for me to understand just how much discipline and focus it would take for me to provide my bees with the most comfortable home.

Today, beekeeping is a successful endeavor in my life which allows me to run a small business selling some of the honey that my bees produce. I could not have made it to this point without all of the knowledge that I have learned along the way. My hope is to share all of my pearls of wisdom and practical advice on starting your own beekeeping operation at home. This book is relevant for those who are seeking to make beekeeping a profitable venture, as well as for those who are interested in the conservation element of beekeeping. Your success in beekeeping deeply matters to me, because I understand how daunting the journey may seem when we are just starting out.

Once you have successfully read this book from cover to cover, you will understand all of the requirements that you need to consider before you can purchase your bees. I will also discuss the time, cost, climate, and space considerations which are important in assessing whether beekeeping is the most suitable hobby or business for you and your family. Thereafter, you will learn a lot more about the secret life of bees and the dynamics of the beehive. This will be of great importance when you begin caring for your bees, because you will understand the common tendencies and behaviors of bees, making your job as the beekeeper more enjoyable and less stressful. I will then explain how to make the most out of this hobby, the organizations and groups that you can become a part of, and the process involved in extracting honey from the beehive and turning it into a delicious source of food and nutritional supplement.

Let us walk together on this journey in becoming better beekeepers. Not only will you richly benefit from what you are about to learn, but the whole species of bees will greatly benefit from your willingness to play a part in protecting their existence. There are so many opportunities that will be opened to you once you learn the fundamental practices and techniques in caring for your bees. What begins as a hobby may soon become a stable source of income. The possibilities are endless for what can come out of your willingness to learn something new. I can assure you that you will not be disappointed by this new skill of beekeeping which you are about to acquire.

Chapter 1

Considerations Before You Start Beekeeping

Beekeeping can be an exhilarating hobby for those who have a genuine interest in learning how to look after bees and harvest honey. Many people see images of successful beekeepers tending to their colonies and they believe that it doesn't take too much to start their own colony. This, however, is not true. The work required to be a successful beekeeper starts before we can even purchase bees. There are a lot of considerations and preparations that we need to be mindful of which set the expectations involved in this activity from the onset.

If we want to reduce the risks associated with handling bees, it is important for us to follow these guidelines. It is possible for us to have a hive and ensure that all of our bees survive and multiply in numbers over time. This goal is not difficult for us to obtain at all. All we need to do is to understand the standard procedures involved in beekeeping and to not deviate from these procedures. Moreover, at the beginning of our beekeeping journey, we must decide on our intentions for keeping bees. If we do not have clear intentions which help us to create a desired vision, we might become discouraged or frustrated within a few months.

I can guarantee you that you will make a few mistakes here and there, and this is perfectly natural. Your mistakes are not confirmation that you should quit or that beekeeping is not suitable for you. Rather, we need to see every hurdle that we might face along our beekeeping journey as an opportunity to learn and improve on our skill. No one is born a master at beekeeping. This hobby requires much practice and a lot of time spent following the same processes until the processes become more refined. Therefore, I urge you to read this chapter thoroughly and understand the implications and requirements of becoming a beekeeper.

What is Your End Goal?

Before we can begin on our beekeeping journey, we must decide on our desired end goal. There are many reasons why people are interested in beekeeping, and these reasons are not limited to producing honey or pollinating garden plants. Some people are drawn by their social responsibility to protect our endangered species, and others have been enticed by the organic products that can be commercialized and sold at local markets and community centers. No matter what the intention is, it is important for everyone to have an end goal before making any purchases. Below are some of the common reasons why people become beekeepers:

1. Pollination

Some people are drawn to beekeeping because of how instrumental bee pollen is for crop production. For instance, bee pollination significantly influences the size, uniformity, and quantity of about one third of our domestic and foreign food crops. Bee pollination also influences other industries indirectly. For instance, it influences the meat industry because animals tend to consume plants which have been pollinated by bees. Therefore, beekeepers may be traders in other industries who would like to positively increase their crop yields through keeping bees in their fields or gardens.

2. Bee Products

One of the more interesting reasons why people are interested in becoming beekeepers is because of their desire to produce products sourced organically from bees. There is a growing demand in the world for organic and raw products in treating illnesses and in restoring our health. Bee products are highly rich in bioactive compounds which carry healing benefits. Some of the properties which are found in bee products have antioxidant, antiviral, anti-inflammatory, wound healing, and anti-bacterial functions. Beekeepers who sell organic bee products can be entrepreneurs, homeopathic healers, or health enthusiasts.

3. Conservation

Some people become beekeepers because of their desire to save honey bees from going into extinction. The frightening truth is that wild honey bees are being wiped out in numbers due to urbanization and its self-serving agenda of cutting down trees and other plant life. Wild honey bees are also threatened by pesticides and parasitic mites which completely devastate these insects. Keeping bees in our backyards may be a sustainable solution for protecting the honey bees from the violence that they succumb to in our environment. Beekeepers who are interested in restoring the lost bee colonies will enjoy the process of watching bee species grow in numbers and fulfill their natural duty of pollinating plants.

4. Harvesting Honey

Many beekeepers come into the field with the intention of harvesting honey in their own backyards. However, not every beekeeper desires to sell their honey. For most hobbyists, the act of bottling your own honey and making use of it at home is enough of a thrill. On the other hand, some beekeepers harvest honey for the purpose of creating their own honey brand and selling it in stores or markets. Once again, there is no right or wrong intention when it comes to harvesting honey. It is up to the beekeeper to decide on the most appropriate path for them to follow. Those who treat beekeeping as merely a hobby are more likely to harvest honey for non-commercial use.

5. Sharing Knowledge

Some beekeepers enjoy working with bees, primarily because they are given an opportunity to study the behavior of these insects and share their knowledge with others. Many schools, organizations, community groups, garden clubs and conservation centers love to network and engage with beekeepers because no one else understands bees as well as they do. Beekeepers may also share their knowledge of bees through literature such as online or newspaper articles, or through books such as this one. Bees are fascinating creatures for so many reasons and to so many people. This presents a great opportunity for beekeepers to become thought leaders and educate the masses on the new discoveries found in how bees operate or in their remarkable social behavior.

6. Bee Therapy

Bee therapy may not be a scientifically proven concept, however, if you ask any beekeeper about it, you will receive a lot of positive testimonies. Many beekeepers have shared how this activity has helped them to reduce the stress in their bodies and become more relaxed. I am certainly one of those beekeepers who believes in bee therapy. Ever since I began my journey, I have experienced a lot of health benefits associated with my beekeeping. Whenever I am working with my bees, it feels as though all of the mental clutter is removed and my mind is focused on the task at hand. The sense of calm is magical and time stands still, even if it is for just a moment. Since I started with beekeeping, I have felt more at one with nature and my attitude toward life is more optimistic and full of gratitude.

Do You Have the Green Light?

Now that you have clarified your intention for becoming a beekeeper, it is necessary for you to be aware of some basic requirements that apply to all beekeepers. As much as we are enthusiastic about keeping bees in our backyards, we need to be mindful of the implications of our beekeeping activities on the community around us. The objective is to care for your bees in peace without your activities being seen as illegal or a disturbance to your community. Therefore, we need to receive the green light on a few factors before we proceed any further.

Legal Restrictions

The first factor which we need to verify is whether we are legally permitted to keep bees in our backyards depending on the area in which we live. Many residential areas permit beekeeping, however each area will have varying requirements regarding the manner in which beekeeping should be handled. This means that we cannot rely on general zoning restrictions which we find on the Internet. We need to check our local zoning laws which will tell us whether or not beekeeping is allowed in our local district. Some zoning districts will permit beekeeping without an individual having to hold a permit, and other districts will require the individual to request a special permit before they can store bees in their yards.

In some areas, it may be difficult to get a hold of the zoning law, or perhaps the law does not specify any guidelines regarding beekeeping. In this case, there are a few people who you can contact who would have some idea on whether or not beekeeping is permitted in that district. For instance, you could contact a local zoning enforcement officer, a representative from a buildings association, a property lawyer, or your local animal control companies. If you live in a condominium or in a property which is part of a homeowner's association, you would need to get in touch with the building supervisor or landlord for more clarity on whether their property bylaws permit beekeeping.

Assuming that your district allows beekeeping as an activity, you would need to find out whether or not there are restrictions or specific requirements regarding the way in which bees should be kept and handled. It is rare for zoning laws to give individuals the liberty to purchase as many beehives as they want and store them in whatever manner they choose. Some of the common regulations that are enforced by zoning laws have to do with minimum lot requirements, the type of hive and bee species permitted, the maximum number of hives, and any fencing or storage specifications. Apart from the zoning regulations, individuals will also need to check whether a registration fee or license is required. Some districts will require individuals to register their beehive with the local zoning office or health board and provide contact details as well as the location of the hive.

Approval from the Neighbors

It's important that we receive the green light from our family, friends, and more especially our neighbors before we go about our beekeeping activities. As much as we may love our little buzzing friends, there may be neighbors who are opposed to being at such close proximity to a swarm of bees. Some communities will require you to notify your neighbors of your desire to keep bees in your backyard. You can imagine the politics involved in getting all of your neighbors to give their stamp of approval. We should expect some of our neighbors to object to our beekeeping activities. Their initial disapproval doesn't mean all is lost; most of the time people will reject bees out of ignorance or misinformation. It is our responsibility to educate our neighbors on the benefits of protecting bees and keeping them in our community.

I find that the most common reason why a neighbor would object to our bee-keeping activities is because they are afraid of being stung. This is a normal fear to have and is valid particularly for neighbors who may be allergic to bees. Once again, it is our responsibility to explain that not every bee is out to sting us. In fact, honey bees will generally only sting a person as part of their defence mechanism in protecting their hive. When it is in the garden pollinating our plants or foraging on fruit or water, it will rarely sting unless unduly provoked or threatened in any way. Therefore, as much as the fear is valid, it is not likely that our neighbors or their animals will be stung by a bee when we follow our safety precautions and they commit to not provoking a bee when they see one.

We also need to explain to our neighbors that they are more likely to be stung by a hornet or a wasp than a bee. However, let's say that our neighbor is furious because they were stung by a bee and the presumption is that it was one of ours. They would need to prove that we were negligent in how we were handling our bees. Their claim would have a better chance of succeeding if indeed we were found to not have followed the legal regulations or made the necessary precautions to keep our hive away from any dwelling place and provide our bees fresh water on a consistent basis. This means that we cannot take any shortcuts in preparing the proper storage for our hives and ensure that we have followed our local requirements for beekeeping.

Making Enough Space

Our hives need enough space in our backyard in order for them to not disturb our neighbors and to give our bees enough room to roam around and work freely. We need to assess how spacious our gardens or fields are for storing our bees.

First, we need to consider the size of a typical hive. The most common type of hive that we find in the market today is the Langstroth beehive, which is a vertical modular structure which has key feature such as vertical frames, a bottom board which allows for the entrance of bees, some boxes which contain frames for brood and honey, and an inner cover and top cap which controls the temperature in the hive and protects it from harsh weather conditions. The dimensions for a Langstroth beehive would typically be 16 by 22 inches.

However, once we invite our buzzing friends into the hive, space becomes a crucial factor. Throughout the day the entrance to the hive will have constant traffic from the hundreds of bees coming in and going out. Beekeepers should expect this traffic to extend across a 5 foot radius around the hive, with a greater concentration of bees swarming at the front of the hive. Therefore, to be on the safe side, I would advise you to allow for a 10 - 15 ft breadth around your apiary (the location where you have decided to keep your bees). I would not recommend that you keep your bees anywhere near your plants or vegetable garden (as aesthetically pleasing as it may look) because no one enjoys having the constant interrogation of bees while they pull out weeds or tend to their crops.

If you don't have a lot of garden space available for your bees, you can do a process known as screening your hive. This would involve constructing a 6ft tall barrier around your hive that would direct the movement of the bees upward (ensuring that you use less space on the ground). Alternatively, city dwellers can make use of their apartment or townhouse rooftops and balconies by placing their hives on them. This strategy is useful in reducing the exposure that the neighbors may have to our bees and ultimately reducing any bee sting incidents. The only inconvenience with placing the hive so high up is that the beekeeper has to travel up elevators or staircases with all of their equipment every time that they tend to their bees. If this is not too much of a burden for you, I believe it would be a great solution for beekeeping in the city!

Cost of Beekeeping

Every hobby comes with a cost. The cost of beekeeping varies depending on how committed you are to this activity. For instance, a hobbyist does not need to purchase a lot of hives because they do not intend to commercialize their beekeeping activities. On the other hand, a beekeeper who is looking to start a small business from their activities would need to invest in quality materials and purchase a lot more hives.

Therefore, to answer the question on the cost related to beekeeping, I would say that it is dependent on one's intentions and level of commitment.

However for cost calculations, beekeepers must budget enough money for the purchase of equipment, tools and accessories, the cost of registering the business (if beekeeping is for commercial purposes), any payable fees to the local authorities, and the cost of maintaining your bees. Generally, for the first year of beekeeping, you can expect to pay roughly between $700 and $1,000. It is important to note that beekeepers will spend the most amount of money in the beginning of their beekeeping journey as they prepare their apiaries and invest in all of the gear and accessories needed for successful beekeeping. After the first year, most of the costs associated with beekeeping will be for the maintenance of your hives (including repairs), medication for your bees, and purchasing more bees.

There are many ways to save costs with beekeeping, such as starting small and only expanding your operations once you have seen the success of your first hive. Another way to reduce expenses is to get handy and build your own beehives. Of course, this would need to be carried out by an individual who is skilled at woodwork, otherwise you will have a compromised hive. Alternatively, you can look at beekeeping as being an investment generating a higher return in the long run. Once your bees are settled in your garden and they have developed a successful routine, you will find that they are fairly low-maintenance, hard workers who will diligently work to return all of your invested income by producing quality honey.

Making Time for Beekeeping

One of the benefits of becoming a beekeeper is that this hobby will not take much of your time. It is estimated that it takes between 15-30 hours a year to take care of one colony of bees. This estimation may be increased during the first year of beekeeping due to all of the preparations that we are required to do. Furthermore, individuals who are looking to make a business out of beekeeping will typically spend more time tending to their bees; generally, they will also have a lot more hives to manage. However, I will say that in order for anyone to master a skill, it is important for them to dedicate a lot of time to practicing their craft.

As I have mentioned before, no one is born with the natural talent of beekeeping. It is a skill which we learn and become better at with time and experience. Even when we're not busy maintaining our hives, it would be beneficial for us to observe our bees and form a relationship with them. Perhaps on a Sunday afternoon when the day is quiet and the phones are not ringing, it would be a wonderful thing to take a stroll in your garden and spend time with your bees. New beekeepers could also take it upon themselves to do extra reading and study up on bees and how to become an expert at their skill. There are many beekeeping clubs, online forums, and subscription magazines which provide meaningful information and current research on the best practices for beekeeping. In other words, I encourage you to immerse yourself in this rewarding craft and see how it adds more purpose to your life and how you choose to live. The more time and energy you dedicate to this activity, the more positive results you will reap from it.

Is Your Personality Suited for Beekeeping?

Similar to any other hobby, beekeeping is fulfilling when we genuinely love and enjoy what we are doing. I can list all of the requirements that are needed in order to become a beekeeper, and they will not deter those who are sincerely drawn to protecting and caring for bees. No one has to convince a beekeeper to go and check on the well-being of their bees, clean the hive, or expand their knowledge on their skill. For these passionate beekeepers, their hobby is a form of service or a way to connect with nature. It is fulfilling even beyond the gift of honey or any other benefit that is offered by the bees. Of course, if we are not passionate about beekeeping, it will be impossible for us to succeed at it and raise colonies of healthy and thriving bees.

Therefore, we must assess whether our personality is well-suited for beekeeping. We need to reflect on our current lifestyles and whether beekeeping would be an extension of how we live or whether it would take a lifestyle adjustment. Those who treat beekeeping as an extension of their existing lifestyle are more likely to succeed because they are already familiar with the conditions related to caring for bees.

For instance, I find that individuals who have a deep love for animals are more suitable to beekeeping. Animal lovers usually prefer the company of their furry friends at the expense of spending time with other human beings. These individuals are also interested in animal rights and activism which creates opportunities for them to care for the well-being of animals and in some cases, adopt stray or vulnerable animals. Moreover, individuals who are nature lovers and enjoy spending most of their time in a garden outdoors are a great fit for beekeeping. I wouldn't say beekeeping is a dirty job, however it does require the use of your hands and not being afraid to clean the frames or move heavy boxes full of honey.

Outdoorsy individuals will find this kind of work challenging yet gratifying because they can feel a sense of accomplishment after a long day of work. Beekeeping may also be suitable for individuals who love gardening or small scale farming. These individuals already spend their time outdoors, and they are lovers of nature and caring for animals. Learning how to care for bees would assist them in improving the quality and health of their gardens or crops. In fact, I believe that these individuals need less of an incentive to become beekeepers than any other person would because there are so many benefits that they would receive from having bees buzzing around in their backyards.

In essence, the presence of bees in any household will provide a significant amount of improvement to the environment of that particular household. However, not every individual is suited for caring and looking after bees. I am so protective over these furry insects because of how valuable they are to our ecosystem. It saddens me when I see individuals storing bees for the purposes of exploiting them for monetary gain. It goes without saying that when the motivation is purely financial, there are a lot of compromises made when it comes to the well-being and health of the bees. Soon the bees begin to die in numbers because they are unable to survive in a stressful environment for long periods of time.

This is why I will always ask new beekeepers of their intentions behind storing bees. All of us as beekeepers have a responsibility to protect what is left of the species of bees and to avoid inflicting any more harm. Therefore, those whose intentions are purely financial are not suitable for the job; this kind of work requires our hearts to be involved and for us to be intuitive to the needs of these wonderful creatures. Furthermore, those whose motivations are purely to harvest honey are not suitable to become beekeepers. There are so many processes involved before honey can be extracted, and it would require more time and effort as opposed to purchasing a delicious bottle of honey at the store. Remember that our primary motive as bee-keepers should be to protect bees and not to exploit them for honey merely because it looks good for us to bottle our own honey at home.

As rewarding as this hobby is, there will be times when you are frustrated because there is a foul smell coming from your hive and you do not know what to do, or perhaps you notice your bees dying but you don't know the cause. In these times, the only reason why we do not throw in the towel is because we have already invested our hearts into our hobby. It would be too painful for us to not continue watching our bees grow and providing a comfortable home for them in our backyards. It is inevitable that you will form an emotional bond with your bees and care for them as if they were a part of your family to some degree. Even when there are hardships or unforeseen costs, beekeepers who are genuinely passionate and committed to their work have the ability to raise healthy colonies of bees and continue their operations for many years.

Chapter 2:
The Life of a Bee

Honey bees, also referred to as Apis Mellifera in the science community, are the most popular species of bee found in America. These flying insects are close relatives to wasps and ants, and can also be found in all continents of the world excluding Antarctica. Honey bees play an indispensable role in economies throughout the world because of their ability to pollinate plant life. When they are not pollinating plants, you will find them slurping up the nectar from deep within flowers and transporting it to their hive where it will be used for honey production. When the bees carry the nectar back home to their hive, their small bodies break down the complex sucrose found in the nectar into two forms of sugar, namely fructose and glucose.

Upon reaching the hive, the honey bee will neatly place the sugar into a honeycomb cell and begin to flap their wings ferociously over the syrupy liquid in order to reduce its moisture and thicken the substance. In order to protect what will later be a source of food for the colony, the honey bee will seal the honeycomb cell with beeswax (see photo below). While one bee is securing food for later consumption, there are other bees which are tasked to complete other household work within the hive. Some bees are hard at work building new honeycomb cells, other bees are taking care of the young larvae, and the largest bee in the hive, who provides the leadership, is moving from cell to cell laying eggs and expanding the colony.

As you can see, honey bees live a highly structured life and they have a level of discipline which most of us would be envious of. These furry creatures also have procedures in place to protect their hive from any external threats such as diseases. The bees make a substance known as propolis which is a combination of beeswax,

honey and tree resin. Propolis is a powerful agent in disinfecting the beehive and it has other antifungal, antibacterial properties. With such a large family consisting of hundreds of bees, good communication skills are vital to the efficiency of every hive. Bees understand that without good communication, their survival is at risk, therefore they have secret signals and social behavior which helps them to send messages to one another and warn other bees of danger.

Bees communicate with one another through scent and, wait for it, dancing. Honey bees have the remarkable ability to release a hormonal scent known as pheromones. This scent changes depending on the message that the bee intends on sharing with the rest of the hive. For instance, when a honey bee is warning the others of an intruder, it will release a warning scent which smells a lot like bananas. However, when the bees are in a celebratory mood and feeling rather happy, they will release a feel-good scent which smells a lot like lemons. Messages are also shared through dancing. For instance, when a worker bee returns home to the hive with nectar or pollen, they will direct the other worker bees on where to find it through dancing. Biologists have coined this dance the "waggle dance" and it consists of a few turns and wiggles to signal where the food may be located. This is not the same way we would share directions with our friends or family, however it proves to be highly effective for bees.

Types of Honey Bees

Honey bees are not created equal. They are organized in terms of the role that they play in the life of the hive. You will never find a bee without an assigned role. In the hive, there is no day off and every bee is expected to fulfill its responsibilities and work as a collective in building a successful colony of bees.

Within the hive, there is a strict social order and bees are placed in three groups. These groups are worker, drone, and the one and only queen. Each hive will consist of a queen, a few hundred drones which are the male bees, and the remainder of the hive will consist of worker bees which are the female bees. Typically, wild hives will contain around 20,000 bees, while contained hives that are managed by beekeepers may contain up to 80,000 bees. Within these large colonies, the mission of the drone bees is to mate with the queen bee, and her mission is to produce eggs and expand the colony. All of the housekeeping tasks such as gathering food, constructing honeycomb cells, and protecting the young bees are performed by the female worker bees. Below is a breakdown of each type of bee and their responsibilities:

The Queen Bee

The queen bee can be identified by her abdomen, which is elongated and extends beyond her folded wings. She is usually the only reproductive female bee in the colony and as I had mentioned above, her sole purpose is to mate and lay eggs. The queen bee usually begins laying eggs in the beginning of spring and will continue for as long as pollen is available (which will typically be until fall). A productive queen bee can lay as much as 2000 eggs per day; we mostly find that the younger queen bees have the ability to lay more eggs than older queen bees. The lifespan of a queen bee can reach up to 5 years, however her period of usefulness to the expansion and power of the hive ranges from between 2 to 3 years.

This is why most beekeepers will replace the queen bee after 2 to 3 years by purchasing a quality queen bee from a reputable supplier. However, in the absence of the queen, the worker bees will assume the role of queen up until a new queen is reinstated. The process of reinstating a queen bee is a matter of urgency for the colony because of the significant role the queen plays. Queen bees release pheromones which help the rest of the colony function effectively and inhibit the development of the worker bees' ovaries. When a queen is removed, the worker bees will notice her absence within a few hours because of the decline of pheromone levels in the hive.

Low pheromone levels in the hive can cause chaos in the structure and organization of bees because it will open the ovaries of worker bees and allow them to lay eggs. Therefore, a new queen must be found, and quickly! Typically an "emergency" queen will be chosen from the youngest larvae available and as time goes on, worker bees will battle for the position of queen. The winner is chosen based on the quantity of the pheromones that she is able to produce. If the bee produces a low quantity of pheromones each day, the worker bees may see this as a sign that the bee is not fit to be the queen and another candidate will be evaluated.

The Worker Bee

Even though worker bees are the smallest in size out of the three types of honey bees, they make up for it with their numbers. The majority of the hive consists of female worker bees who typically do not have any reproductive capabilities (these capabilities are reserved for the queen). Worker bees are the ones responsible for doing all of the household and administrative tasks within a colony of bees. Some of the invaluable tasks performed by this type of bee are secreting beeswax and moulding it into honeycomb cells, searching for nectar and pollen in the surrounding environment and bringing it back into the hive to transform into honey, producing a substance known as royal jelly to feed to the queen and the younger bees, cleaning the hive of dirt and removing dead bees, protecting the hive against external threats, and regulating the temperature conditions within the hive.

The lifespan of a worker bee is a lot shorter than the lifespan of the queen; these bees tend to live for up to six weeks, and the first two weeks of their lives are spent doing tasks in the hive. Worker bees that mature are able to live up until the next spring. In the winter months, their duties involve clustering in numbers around the queen bee in order to keep her warm during the cold months, and once springtime comes they will raise and nurture the new larvae. It is difficult to imagine how a colony of bees would function without the worker bees. Their support and manual work is what helps to keep life flowing in the hive and the new generation of bees develop in a well-organized and healthy environment, with plenty of sweet food for nourishment.

The Drone Bee

Drone bees are the male bees in a colony, and their sole purpose is to mate with the queen and reproduce new life in the hive. We can identify a drone bee by their size; they tend to be fairly larger than worker bees and possess large striking eyes which are positioned on the top of their heads. Due to the fact that they do not gather food, the drone bee's mouth will be significantly reduced in length to that of the worker bee. These bees are significantly less in number than the worker bees because they are the product of unfertilized eggs which were laid by the worker bees (in the absence of the queen).

Therefore, beekeepers are more likely to find new drone bees in the hive during spring and summer, usually four weeks before a new queen is reinstated. This period of reproduction is also useful for the colony because when a lot of drone bees are developed, it ensures that there will be a vast number of bees to mate with the new queen when she assumes her role. The production of drones declines around late summer when the amount of available food starts to dwindle. As winter starts approaching, the worker bees in their protective mode begin to drive the drone bees out of the hive out of fear of food scarcity should the drones continue to mate with the queen. However, when there is no queen, the worker bees will start to lay eggs and produce drone bees in significant proportions. This spells the beginning of doom for a hive because it will be difficult for the colony to survive with the scarcity of food during winter months. If the beekeeper is aware of this debacle, they are able to purchase a new queen from a bee supplier and save the hive from imminent danger.

The Life Cycle of a Honey Bee

Every bee has a life cycle which spans between a number of weeks to a number of years. Before each bee assumes its role in one of the three social orders, it will follow a standard process of growth from when it is an egg until it is an adult and assumes adult responsibilities. Once again, this growing phase will vary among the bees, but the general time frame will be 24 days for drone bees, 21 days for worker bees, and 16 days for queens.

The journey of a honey bee begins as an egg which is laid by the queen bee. As a beekeeper it will be important for you to know how to spot eggs (resembling minute grains of rice) because the presence of eggs in a hive is one of the definitive signs that the queen of the hive is still alive. Developing the skill of spotting eggs requires time and practice, however the more time you spend observing your bees, the easier it will be to spot these tiny eggs. Let's see if you can spot the eggs in the photo below:

The queen lays her eggs in a single cell which has already been cleaned and prepared for the new deliveries by the worker bees. If by any chance the queen finds the cell with a speck of dirt, she will move on to another clean cell where she can safely lay her egg. It is interesting to note that the cells are different sizes and the size of each cell determines the future of the bee; whether the bee will be a worker or a drone. For instance, if the queen releases a fertilized egg into a cell fit for a worker bee (smaller in size), then the bee will be groomed to become a worker. Wider cells are usually drone cells and when the queen stumbles upon them, she will release an unfertilized egg in each drone cell. It is the responsibility of the worker bees, who construct these cells, to determine the ratio of female to male cells. Therefore, if we look at a honeycomb cell, we will find a lot more smaller (female) cells than wider (male) cells.

Three days after the eggs were laid in the cells by the queen bee, they hatch and become larvae. We can identify larvae by their snowy white color, resembling small little grubs in their individual cells. At first the larva is tiny, however it quickly grows and sheds layers of its top coat five times. Worker bees are kept busy when larvae start to appear because they demand a lot of food and are known to consume 1,300 meals each day. Worker bees will feed the larvae royal jelly at the beginning of their development, however once they grow older they will be fed a mixture consisting of honey and pollen. It only takes around five days for the larvae to grow over 1,570 times their original size and it is usually at this point where the worker bee will cover the larvae with a permeable seal of beeswax. Once they have been sealed in, the larvae will spin a cocoon around their bodies.

At this stage, the larva is now a pupa. This is the stage where we begin to see the body of the bee taking shape and becoming more defined. Of course, the beekeeper cannot see all of this moulding happening because the transformation is hidden beneath the layer of beeswax. At the pupa stage, the eyes, wings, and legs begin to develop and the bee starts to resemble a grown adult bee. The coloration begins in the eyes, and the color will develop from pink, to purple, and then finally turn black. After the bee has received its full color throughout its body, it will start to develop the hairs which will cover its entire body. The pupa stage takes around 12 days and once this period is over, the now adult bee will free itself from the wax seal by chewing it until the bee is fully exposed. Now an adult bee, it can join the rest of its family and assume its role in the colony.

Bee Pollination

Perhaps one of the most renowned gifts that bees provide to all of humanity is the ability to pollinate crops and other plant life. Plants cannot reproduce or grow successfully without pollination. Therefore, without the existence of bees (and other animals which pollinate) our global agricultural industry and all plant-based foods and materials would be compromised. The bee is not at all conscious of the significant role that it plays to our economy; in fact, its ability to pollinate happens by accident. When a bee approaches a flower to collect nectar and pollen for its hive, some of the pollen from the stamens (this is the male reproductive organ of a flower) gets caught in the hairs of a worker bee's body. When the bee moves on to another flower, some of this pollen on its body is rubbed off onto the stigma (this is the female reproductive organ of a flower).

As soon as this process takes place, fertilization in the plant is possible and it can now produce its fruits. Therefore, it is quite clear for us to see just how much plants rely on the visitation of bees and other insects and animals in order for them to reproduce. Over time, plants have adapted to become more attractive to bees, however these furry creatures still have favorites.

Bees are more likely to buzz around plants which have flat or open flowers with plenty of nectar and pollen. Furthermore, a bee can be drawn to a flower by its particular scent which is very appealing to the worker bee. If the structure of the plant and the fragrance of the plant cannot convince the bee, the plant can also attract bees with its distinct and vivid color.

Although a honey bee's main diet is nectar and pollen, it is able to collect other juices and liquids found in plants and fruits. Sometimes, when the honey bee finds a plant bug which secretes honeydew, it can collect this sugary liquid waste and store it as honey. In some cases however, honey bees cannot seem to find pollen, honeydew, or nectar at all. When they are presented with this dilemma, the bee will decide to collect plant spores or any amount of animal feed that they can find and store them in the hive as they would the other sources of food. It is a sad thought to imagine hungry honey bees working very hard to find food in what remains of our natural environment.

Beekeepers assist honey bees by taking away the stress related to sourcing food because the bees have ample amounts of food that they can get a hold of all year round in the backyard of the beekeeper. Beekeepers understand that honey bees require a lot of nutritional supplements that are found in plants including the vitamins, minerals, proteins, carbohydrates, and lipids. These nutrients are collected to feed every bee in the colony, from the larvae to the queen. Without the necessary nutrients, the new generation of bees cannot survive and the overall performance and health of the colony will decline. Therefore, the role of a beekeeper in protecting the life of a colony of bees is priceless. With access to the right conditions and an environment full of nutritional food supplies, species of bees are able to continue growing and reproducing strong bees.

Hey, did you know...

- Approximately one third of all the food that we consume is a result of bee pollination.

- During its 6-8 weeks of life, a worker bee would have flown a distance equivalent to one and a half times the circumference of the Earth.

- Bees are drawn to the smell of caffeine.

- A worker bee will visit 50 to 100 flowers on a single trip to collect nectar and pollen.

- The famous "buzz" sound that honey bees make is generated by their wings beating 200 times per second.

- In order for worker bees to produce a pound of honey, they must fly approximately 55,000 miles.

- Honey bees do not sleep. Instead, they become motionless in order to preserve their energy for the following day's work.

- Worker bees are not born knowing how to produce honey. They are trained from a young age by the more experienced bees, and soon they become expert honey producers.

- Worker bees will die after they have used their stinger, however a queen bee will survive even after it has stung an animal with thicker skin because of its smoother stinger. However, it is rare for a queen bee to sting a human (it will reserve its stinging for battles with other queen bees).

- The term "honeymoon" comes from an ancient ritual where a newly married couple would be supplied with a month's worth of mead (fermented honey drink) in order to bless the union with happiness and fertility.

Chapter 3:
Starting Your Beekeeping Hobby

I can clearly remember the first time that I performed my first hive inspection. I could feel my heart pulsating louder and louder. The truth is that I was both excited to see the progress of my bees and also terrified of being stung! My goal was to lift the frames as gently as possible without disturbing the peace of the bees or causing any suspicion within the hive. Fortunately, I was able to complete my inspection successfully and all of my fears soon faded away. This was the first moment beekeeping became real for me; it was no longer a concept or a noble service anymore. I felt responsible for these little guys, and my hobby soon became part of my lifestyle.

It is difficult to not fall in love with beekeeping as soon as you get a taste of it. This humble hobby always reminds me of the simplicity of life growing up; we had a lot of respect for nature and animals because we saw the value that they brought in our home. Nowadays, we need to make a conscious effort to connect with nature or find groups of people who share a similar love because our society is moving in an opposite direction. Beekeeping as a hobby is one of the ways that we can reconnect with nature and find ourselves while tending to our insect friends. For someone to understand how genuinely rewarding this hobby is, they would have to experience it first hand. I cannot put it in words how liberating it feels to be out in the yard and playing a crucial role in the life of a hive.

The success of a hive feels like a personal success because we know how much we have had to invest to ensure that these hard working insects get to experience the most of their short lives. However, as rewarding as beekeeping is, it requires a lot of prior planning and preparations for it to be prosperous. Beekeepers need to be aware of the reality of colony collapse and to have strategies put in place to reduce the risk of losing hundreds of their bees over a short period. We are required to continue educating ourselves on the risks that our bees are exposed to, especially in the community in which we live and to also network with other beekeepers who can share tips and mentor us on our beekeeping journey.

However, even after receiving a mouthful of advice from multiple sources, it will not completely prepare us for this unique experience of caring for bees. I believe that each beekeeper needs to build confidence in themselves and understand that they are the only ones who truly understand their bees. You are the one who spends time with your bees every day and assess their behaviors. You are the one who can sense when something is just not right in the hive or when the bees are not happy. You must learn to trust your own gut and realize that you are perfect for the job. At times you may ask yourself, "Am I doing the best for my bees?" And the truth is that you are! The beginning of any pursuit is rocky, but eventually you reach a place where you are familiar with the procedures and expectations and suddenly the work is pleasurable. This section of the book will prepare you as much as possible for a smooth beginning as you prepare to fully embrace this hobby.

The Best Time to Start and How to Choose Your Bees

The best time to start beekeeping is in the springtime, however it does vary month to month depending on the climate. The general rule is to begin your operations when the cold or rainy season comes to an end and the flowers are beginning to blossom. Nonetheless, the beekeepers' work does not begin in spring. We need to prepare well in advance, usually starting our preparations and gone of the athering our equipment in autumn. Beekeepers need to also gauge the climate in their environment to get a clear indication of when they should move in the bees. For instance, in America there are different climates between the east and west coast.

On the East coast, the environment is usually hot and humid in summer and cold and dry in winter. The west coast experiences a slightly different climate; for instance in the southwestern region, it is typically warm to hot throughout the entire year. Researching the climate in your region will form part of your information gathering stage coupled with the investigation on your region's laws on beekeeping. Once you are familiar with your climate, it will be easier for you to plan the ordering of your bees. The process of ordering bees must be done no later than January or February. This is so that we can ensure that our bees are shipped to us by April.

I find that planning to order bees in advance reduces the risk of receiving my bees too late, the supplier running out of stock, or having very little time to get the bees accustomed to their new home. Usually when the bees arrive a little earlier than April, the beekeeper will have to feed them up until they are able to go out of their hive and search for their own food. During the hot summer months, it is important for the beekeeper to make sure that their bees have fresh water accessible to them on a regular basis, because the heat and humidity in the air can be life-threatening to the hive. The worker bees will collect the water and spit it inside their hive in order to cool down the temperature. Keeping a water source close to the hive will also protect your neighbors from receiving unwanted visitors looking for a refreshing drink.

Bee Hive Location

It is important for us to consider where to place our bees once they arrive at our homes. The location of our bees is important for their safety and success as well as for the safety of our neighbors and community at large. As I have discussed before, receiving the green light from the local authorities and our neighbors should be our first priority for all the reasons that have been mentioned. Once we have received their approval and support, we must consider the most appropriate position to place our hives in our backyards. An important note to remember when picking the right position is that bees need to have a safe flight path which directs the incoming and outgoing traffic in the hive. Therefore, we should face the opening of our hive in the direction that we desire most of our bees to fly. This direction should be away from our property or our neighbors' yards and in a direction where there isn't much disturbance.

The perfect position for our hive is also an area which receives partial sunlight. This means that it is not directly exposed to the sun because our hive must remain cool. If this setup is not possible, we can place our bees in a sunny area in the morning when they need to wake up and become fully active, however in the afternoons we must move the hive to a shady spot so that the hive can be cooled by the afternoon breeze. Moreover, the perfect position must be one which is not too far away from the designated area where we intend on harvesting our honey. This is because the honey supers (the boxes storing honeycombs) become very heavy when they are full of honey. We need to plan ahead of time for a way to haul these supers home and begin the process of removing the honeycombs.

Beekeepers need to also consider drainage options to ensure that the hive does not become filled with moisture or drowned in a flood of water. One of the ways to reduce moisture is to avoid placing the hive in a cold shady area for long periods of time without any exposure to sun. It may also be a good idea to place the hive on top of bricks or an outdoor table so that it is suspended from the ground. This will reduce the likelihood of our hive flooding, however it will also limit the number of intruders which enter the hive (unwanted bugs or weeds from the ground). Hives should also be placed in a fenced area where they are protected from animals or human beings who may vandalize the hive and threaten our bees.

Choosing the Perfect Bees

After having done the necessary preparations at home and received the green light from our local authorities and wonderful neighbors, it is time to choose which species of honeybee we are interested in purchasing and making our order. This is such an exciting time for a beekeeper because it is the moment before our new friends come home! Nonetheless, our excitement shouldn't distract us from the many decisions we need to make about what kind of bees we are looking to keep and how we desire to store the bees once they have arrived.

The first order of business is to decide on how you intend to source the bees. Bees can be sourced in the wild by trapping a wild swarm of bees, or they can be sourced from an apiary or commercial breeder. If you are going to choose the supplier route, it is important to make sure that you've done your research on the supplier and have seen great reviews from other beekeepers on the quality of the bees that they sell. Trapping a wild swarm of bees is a difficult and dangerous procedure, however it does ensure that the colony of bees is locally sourced with a higher probability of survival.

Nevertheless, I find that the general trend of beekeepers is to source bees from an accredited bee supplier.

Once we have chosen our supplier and are ready to make our purchase, we need to make another important choice between purchasing our bees in a package or in a nuc, which is the shortened word for nucleus colony. Even though it is much more convenient to purchase our bees in a package or nuc, we also accept the risk of having bees which are not locally sourced. It is difficult to see whether the bees are local or not in the beginning, however as time goes by it will be evident in how the bees manage to adjust and thrive in the environment in which we place them. One again, we cannot take any shortcuts in sourcing our bees; we must ensure that they are locally sourced even if it means paying an extra amount of money.

Packages vs Nucs

When purchasing bee packages, we need to understand that the queen will not be related to the other bees because apiaries generally breed queens separately from drone and worker bees. When the apiary staff are preparing a package (to be picked up or shipped) they will first collect the queen bee and place her in a cage alone. This is because the rest of the bees will consider her to be an intruder until they have become more familiar with her pheromones and can welcome her to their family.

Once the queen has been caged, the staff will proceed to the bee yard and collect a ton of bees and place them in another cage. The only issue with this process of mindlessly dumping bees into a package is that the staff member does not consider the ratio between worker bee and drone bee. Bees are typically sold in 3 to 5 pound packages containing between 10,000 to 12,000 bees, which is an adequate quantity to begin your beekeeping hobby.

After having purchased a package of bees, it's time to transfer them into the hive that you've purchased which will become their new home. You can expect the shipped package to include the bees in their separate cages as well as a can of feeder syrup. Do not be shocked if you see the bees swarming around the can. In fact, if you're considering calming them down before you open the package, it would be a great idea to spray some smoke through the side of the package which will make them retreat and give you some room to maneuver quickly. You will find that underneath the lid of the package, there will be the top of the feeder can and the piece of metal attaching the queen's cage to the rest of the package.

For your convenience, you can take a small pry bar and carefully remove the feeder can without allowing the queen cage to fall in. Once you have placed the feeder can aside, quickly replace the lid back on the package. Now you can safely pull out the queen cage and replace the lid when it has been placed aside. Notice a small piece of cork located on one side of the cage. Remove this cork using an ordinary corkscrew that you would open a bottle of wine with. You should now see what looks like sugar candy appearing underneath. This sugar candy is what separates the queen bee from the rest of her future family. As the two become more familiar with each other, the bees will chew the sugar candy until she is finally released and joined together with the rest of the colony.

Now it is time to insert your bees into the hive. Begin by placing the queen cage into the hive. Next, proceed by removing enough frames or bars in order to create space to dump the bees in. Once the space has been made, you can quickly smoke the bees so that they are less agitated when they are released. When you're ready, open their cage and "yank" them out, allowing them to fall into the hive.

After five days have passed, visit the bees and perform a hive inspection to check and see how they're settling in. The queen should be released at this stage, however if she has not been released yet, you can free her by removing the candy or alternatively pulling the wire mesh off the front end of her cage.

Nucs, on the other hand, are mini colonies consisting of frames, a queen bee, and the natural number and ratio of worker bees and drone bees. Nucs are a great start-pack for new beekeepers who may be overwhelmed with having to organize bees into a hive and create a prospering hive culture from scratch. Within nucs, we will find a family of bees already comfortable with each other and having already started creating the natural resources we would find in a hive. The only disadvantage to purchasing a nuc is that most of the training and feeding of the bees happens under the control of the apiary and it therefore takes away the experience of bonding with the colony or influencing how the hive is managed.

Installing a nuc into a hive is a lot less demanding than installing bees coming from a package. One of the first things you need to check is that you purchase a nuc which has the same number of frames (as well as the same depth) as your hive. Once this has been checked, remove the empty frames in your hive, and replace them with the nuc frames, making sure that you place each frame in the same order that it was removed from the nuc. This is because the bees may have started producing wax and we do not want to cause any confusion.

Types of Bees

Many people are unaware that there are different races of honey bees. Each race or species has its own unique characteristics that may or may not be compatible with the beekeeper's expectations. It's always a great idea to ask your bee supplier of the type of bees that they offer before accepting any package of bees. It always helps when you have done your research prior to purchasing bees, so that you can arrive at the apiary with all of your questions and concerns. However, every beekeeper has a bee suitable for their backyard, and finding this type of bee will make your experience in beekeeping more enjoyable.

The most popular bee in North America is the Golden Italian, scientifically known as apis mellifera ligustica. This type of bee is also recommended for all beginners. Besides having a beautiful physique, these bees are known to be very gentle and extremely hard workers. With the Golden Italian, you will have less incidences of sting attacks because they do not have swarming tendencies. They are also very independent and have the ability to clean their own hives and take good care of themselves. Perhaps the one thing which is disadvantageous about this type of honey bee is that they can wander off and abandon their hives to join another family of bees. They are also known to not travel too far looking for food, therefore the beekeeper should have plenty of plants for them to indulge in at a close proximity to the hive.

Beekeepers who live in rural areas with an unpredictable amount of food available for the bees will find the Carnolian honey bee, scientifically known as apis mellifera carnica, to be the perfect fit. This bee is gentle like its Italian cousin, however it also has a level of resistance to pests, making it suitable in the wild. Moreover, the Carnolian has the ability to build their colony fast and regulate the population in the hive depending on the amount of food which it has available. This bee is adaptable to extreme natural environments where food scarcity or extreme weather conditions would be a factor in their survival in the hive. They are able to work longer hours than the other bee species and travel farther in search of food. The only downfall to this bee is that it swarms easily and it struggles to cope with extreme heat (it usually prefers colder and wetter weather).

The Caucasian, scientifically known as apis mellifera caucasica, is a honey bee species recommended for beekeepers who have a lot of experience taking care of bees. This is because these bees tend to be high maintenance. The great part about the Caucasian is that it can make a lot of propolis. This is good because propolis works as an antiseptic for the hive, however propolis is also a very sticky substance which means more cleaning for the beekeeper.

They are productive bees and work very hard during the summer, although at times they can be excitable or agitated which can quickly become dangerous for any individual crossing their path. These bees also tend to require frequent medical assistance because they are more susceptible to infections than any other species of honey bee.

For the survivalist beekeeper who prefers to catch a swarm of bees from their area, feral bees may be the most appropriate for you. Firstly, these bees are a mixed breed or combination of the other species. They are free-spirited bees that anyone can find wandering among plants and trees. Since they do not belong to anyone, beekeepers can catch them and keep them in their backyards for free. Nevertheless, the fact that the Feral bee is a mixed breed means that its characteristics are unpredictable. The beekeeper can either have a sweet and gentle little friend or an aggressive bee from hell. The only benefit is that they are adapted to the environment and so they have a higher chance of survival in the hive.

Lastly, beekeepers who are also gardeners or subsistence farmers may be drawn to the Mason bee, scientifically known as osmia lignaria. Even though these bees do not produce honey or beeswax, they are perfect for pollinating vegetable gardens or fruit trees. Mason bees are also known to not sting, therefore they are safe to have around children and other animals. I find that these bees are also suitable for beekeepers who are urban dwellers because they do not require a lot of yard space in order to thrive. You can purchase a Mason beehive which can hang on a tree or a fence, saving you a lot of garden space!

Bee Supplies

So, we have now placed an order for our bees. This is a great milestone, however we need to purchase bee supplies before we can do a victory dance. These are the supplies that we will utilize every time we interact with our bees. Don't see them as "nice to have" items; these supplies are a necessity. The advantage of purchasing supplies at the beginning of your beekeeping journey is that you will never have to purchase them again unless they tear or become worn out. I remember receiving a package with all my supplies neatly packed inside. As I opened the box and organized all of the tools and accessories in my garage, I felt like I was inaugurated as a beekeeper. For me, having all of the necessary supplies made me feel more prepared to immerse myself in this amazing hobby. Looking back, I do not regret investing in these supplies at all. Below, I have listed all of the supplies that you will need as you prepare to become a beekeeper.

1. Beehive

It goes without saying that you will need a beehive. This is the wooden house that your bees will call home. You have options when it comes to purchasing your hive. You can choose to purchase a hive kit and reassemble it by yourself at home. The kit will provide you with the individual components that make up a hive, namely the boxes, frames, and the foundation. Alternatively, you can purchase a fully built hive assembled at the factory warehouse and shipped to your home. There are three types of beehives available for you to choose from: a langstroth, horizontal top bar, or a warre hive.

Langstroth hives are the most popular type of hive for many beekeepers. When we see images of white boxes stacked on top of each other, most of the time we are looking at a langstroth hive. These hives may range in the number of frames that they consist of; for instance, we can purchase an eight-frame langstroth hive or a ten-frame langstroth hive. Of course, the fewer frames we have, the lighter our box will be to carry and the fewer number of bees it will hold inside. This vertical tower of boxes is topped with a protective hood necessary for ventilation for our bees. The boxes at the foot of the hive are where the young bees are raised and where the queen bee tends to live; the boxes on the upper level of the hive are used for the production of honey and beeswax. The only downfall of a langstroth hive may be its weight, which makes it very heavy to carry around the garden, especially when it is full of honey.

The top bar hive is becoming more and more popular among beekeeping hobbyists and sustainable farmers. The design of the hive is trough-shaped and consists of a range of horizontal bars protected by a removable top covering. This type of hive seeks to mimic how bees would naturally build their hive in the wild. For instance, the horizontal bars allow the bees to make their honeycomb downwards from the bars, making this a very natural process. In this type of hive, you will not find frames or a foundation because the bars are able to separate the colony while giving enough room for the honeycomb to be built. This makes the work of the beekeeper less intrusive for the bees and usually the beekeeper does not need to use a smoker to disperse the bees. However, the disadvantage with using this hive is that sometimes the comb may break or fall if it is not hung securely on the bars.

The warre hive is sometimes referred to as a top bar hive built vertically. This is because the warre hive consists of a stack of small boxes that carry top bars rather than frames inside of them. Similarly to the top bar hive, there are no frames nor is there a foundation in this hive. The honeycomb is built from the top bars downward and more bars can be added in the lower levels of the hive should more space be required. One interesting element about the warre hive is that it includes a quilt filled with wood shavings or sawdust and an angled hood (which resembles a house roof) which is used for ventilation.

This design provides better climate control in the hive because the quilt will absorb excess moisture and it can be released through the roof. The disadvantage with utilizing this hive is that the top bars cannot be removed for inspection or cleaning. This means that most of the time, the beekeeper will not know how their bees are doing inside the hive. It is only when the beekeeper collects honey from the upper boxes that they are able to remove the honeycomb and check on the well-being of the bees.

2. Honey Bees

Our list of supplies must include the stars of the show: our choice of honey bees. As I have mentioned above, beekeepers have the option of picking a species of honey bee that is more compatible with their lifestyle and environment.

3. Protective Gear

I believe all beekeepers should invest in protective gear because it reduces our likelihood of being stung by a bee. I would be misleading you if I told you that you will never get stung by a bee, however I can promise you that the chances of you being stung are significantly reduced when your body is covered head to toe. Firstly, you will need a veil which covers your face and neck region from being attacked by bees. I always advise other beekeepers to at least wear a veil when they are working with a hive, even when they have purchased a top bar hive. Our body can handle bee stings fairly well in any other part except our faces, which are very sensitive to stings (especially near the eyes, ears, and nose).

Next we will need to wear safety gloves. Experienced beekeepers can comfortably work without gloves, however these are compulsory for all beginners to wear. Our exposed hands are more likely to be stung by bees when we are lifting and fitting frames in the hive. In some cases, the stinger can penetrate the glove, however it will feel a lot less uncomfortable through our thick protective layer. We can use any type of glove, from dishwashing gloves to professional beekeeping gloves that reach up to our elbow, covering our hands and arms. The most important tip to remember when choosing gloves is to pick a pair that will give you a firm grip and minimize the risk of any clumsy accident such as dropping a frame to the ground (which will cause an unnecessary riot).

Lastly, we need to purchase a bee suit which can be a two-piece or a one-piece bodysuit that covers our bodies from the neck downwards. I would say that this is the only piece of protective gear that is optional. If you do not have a bee suit, you can wear clothing which fully covers your body, avoiding textured fabrics which create an odor that aggravates the bees. It is advised that you wear clothing that is dull or neutral in color so that you do not draw any unwarranted attention to yourself. If possible, you should also ensure that your clothes are tight around your wrists and ankles so that you do not have any curious bees exploring your anatomy. Below is a photo of a beekeeper tending to their hive dressed in full protective gear. Imagine yourself dressed in your full protective gear, because it will happen very soon!

4. Bee Tools

There are quite a few hand tools that are sold online or in gardening stores which make looking after bees a lot more pleasurable. The hive tool, for example, is similar to a crowbar and its job is to help us separate the stacked boxes which are sometimes glued together with beeswax or propolis. A scraper is also a great tool for helping us scrape off wax or sticky propolis which has built-up on the boxes or around our frames. We can also purchase an uncapping scratcher which is useful when we want to collect honeycomb from our frames. This tool helps us get between the mesh and effectively uncap our comb. Lastly, we cannot forget about the handy smoker which I would recommend for all beginner beekeepers to have. This piece of equipment helps us calm the bees before we open the hive and conduct our inspections or gather the honeycombs.

5. Honey Extractor

The honey extractor is a useful device for beekeepers to use when harvesting honey. The main purpose of this device is to extract the honey from our honeycombs. I would recommend for hobbyists to lease this device only when they need to use it. On the other hand, beekeepers who intend on producing honey for commercial purposes should purchase a honey extractor and keep it as part of their inventory. Beekeepers have a choice between manual hand-operated extractors and those that are operated by a motor. New beekeepers are advised to rent one out for the first few seasons so that they can reduce operational costs.

6. Bee Food

Our bees depend on us as beekeepers to feed them for the first year or so after entering the hive or until they are able to produce their own honeycomb. Bees will make producing comb one of their first priorities because it is useful in providing them with a place to live, lay eggs, and store food, as well as a place for brood rearing. The process of building a comb is slow and in order to make it progress faster, bees need a source of food for energy. Therefore, our help will go a long way in securing the survival of the hive. Every beekeeper will have their own cut-off date where they will no longer feed the hive; sometimes this cut-off date is also influenced by budget constraints and the energy it takes to constantly feed the bees.

Nonetheless, we need to remember that new colonies need all of the help that they can receive from us, and so I encourage new beekeepers to continue feeding the bees for as long as it is feasible to, or until you start seeing evidence of honeycombs. Alternatively, beekeepers may purchase nucs which have combs already starting to form. In any case, all beekeepers must be prepared to feed their growing colony of bees with a pollen substitute and/or sugar syrup in order to give them the necessary boost of health.

When feeding the bees sugar syrup, beekeepers must first remove the honey supers so that the bees do not think that the syrup must be stored and used as one of the ingredients to make honey. To feed the bees, you can purchase a bee feeder (there are many for you to choose from in the market). Once you have chosen your feeder, the goal is to make sure that it is constantly full of sugar syrup. Other appropriate foods to feed bees are honey, pollen patties (these can be placed on top of bars), or fondant, which can be fed to bees in winter when sugar syrup might be too cold for them to consume.

Chapter 4:
Caring For Your Colony

When your bees are safely in their hive, you can release your first sigh of relief. Congratulations, you have completed the most grueling part of the beekeeping journey, and you should be proud of yourself! Your bees are now developing a routine and learning how to become more self-sufficient. Of course, you will still need to feed the little ones for a little while longer before you can step back and allow them to thrive on their own. In this section I want us to discuss ways that we can care for our bee colonies and ensure that our hives survive for years to come. Even though the bees will be doing most of the work, beekeepers are still required to maintain an environment that supports the development of the colony.

Beekeeping Tasks By The Season

One of the reasons why I love taking care of bees is that they are so low-maintenance. By this I mean that bees tend to follow a seasonal pattern and as long as we follow this rhythm, we do not have to perform any unexpected tasks. This provides us with a lot of flexibility in dedicating our time to beekeeping throughout the year. All we need to do is to ensure that in every season of the year, we are fulfilling our expected beekeeping tasks. As the seasons change our bees will require our help in different ways and our responsibility is to be there when they need us and thereafter continue with our personal lives.

Spring (March to May)

Springtime is a busy time for beekeepers. This is the time when most will purchase bees and set up their beehives. Since most beekeepers are likely to have new bees arriving at their apiaries, they are advised to focus on feeding their honey bees sufficient food, in order for them to survive the first few months of living in the hive. Essentially, beekeepers will have pounds of sugar stored in their homes to prepare for the feeding months which end as soon as the flowers start to open up and blossom. In spring, a colony of bees can expand quite rapidly, and soon there is not enough space in the hive to accommodate the growing numbers.

When bees are cramped in a small hive, their solution is to swarm. Swarming is the process of a colony splitting in two. As beekeepers, we do not want some of our bees having to relocate to another environment because of a lack of space. Therefore, we must ensure that our hives can be adjusted to include more boxes for the bees to live, otherwise we should position one or two empty hives in close proximity to our active hive so that some of the bees have a new home to live in within our yard.

By April, beekeepers can begin harvesting honey, prioritizing any honeycomb that was not used during the winter months. In order for a beekeeper to keep up with the honey supply, they will need to add more boxes to the hive, giving the bees more space to build their combs.

Spring is also the best time to conduct hive inspections. It is necessary for all beekeepers to form a relationship with their bees. With experience, beekeepers will be able to analyze the social behavior of their bees and intuitively understand their needs. However, in the beginning, there are a few factors that we need to inspect in our hives which will confirm the health of our bees. For instance, a beekeeper may inspect the condition and quality of the honeycomb, looking for eggs within the cells, signs of pollen, and any hive pests or signs of diseases.

Summer (June to August)

June is a delightful time in the hive because the bees are active and the hive is progressively growing stronger every day. In this time, the beekeeper may want to inspect the hive and assess the health of the queen and whether she is still laying eggs. If everything is going well, the beekeeper might like to add more honey supers so that they can collect more honey from the bees. Due to the hot and humid summer sun, it would also be wise for the beekeeper to ensure that fresh water is always available as close to the hive as possible. The position of the hive can also be assessed to ensure that it is receiving an adequate amount of shade and sun.

Summer is also a great time to inspect the hive for diseases and protect it against any intruders. Combs that are diseased or are not hanging correctly on the bars will begin to show. Remove any defective combs from the hive, dead bees, and keep a watchful eye for any Varroa mite infestations. Beekeepers should also reduce the entrance of the hive and make it difficult for wasps and yellow jackets to enter the hive. Reducing the entrance of the hive will also protect weaker hives from being robbed of nectar and pollen from bees coming from stronger hives. The bees will have a better ability to defend their hive when the entrance is reduced and fortified by combative bees.

Autumn (September to November)

September is usually the peak for honey harvesting and beekeepers have a few months left for collecting as much honey as they desire. However, this time is also significant because both bees and beekeepers are preparing for the winter months ahead. This means that all of the tasks that we perform in autumn are to prepare for hive survival during winter. Therefore, as much as we are excited to gather honey, we should refrain from taking too much of it because it will be a source of food for the bees during the winter months. We should also check the quality and health of the honeycombs and discard any diseased combs. Beekeepers can also install mouse guards at the entrance of the hive and further reduce the opening to the hive.

Autumn is also a good time to treat the hive and disinfect it. To treat their hives, beekeepers will first remove the honey supers and then spray the entire hive with chemical mite control. Moreover, beekeepers must inspect the ventilation system in the hive and ensure that it is functioning appropriately and removing excess moisture. Beekeepers will also assess the overall quality of each hive and combine weaker hives with stronger ones in order to help them survive the harsh weather conditions during winter. If there seems to be an inadequate amount of comb reserves in the hive, beekeepers must begin to feed their bees sugar syrup through the feeders (this will usually begin around October) and continue feeding them throughout the winter months.

Winter (December to February)

Winter can be a stressful time for bees if they have not made the necessary preparations in the fall. It can also be a nerve-wracking time for the beekeeper because they are not able to spend as much time handling the hive and conducting inspections as they did in the warmer months. However, if the beekeeper was diligent in completing all of the disease treatments during the fall, they do not have to worry about any external threats disturbing the peace in the hive. The only duty of the beekeeper is to make sure that the hive is protected against harsh winds or floods.

For instance, we can place a brick or heavy weight on the top of our hive to protect the tops from flying away in the wind. It would also be a great time to relocate our hive and place it in a position where it is not in direct contact with the harsh elements in our environment. We should also remember that our little friends are tougher than we think and when we have done our part, we can rest assured knowing that they will continue caring for themselves. In the meantime, we can keep ourselves busy during the winter months by continuing to upskill ourselves and find new ways of making beekeeping more meaningful in our lives. We can also plan to purchase new equipment or expand our bee operations in the spring.

Protecting Yourself

All beekeepers, whether they are hobbyists or entrepreneurs, have to face the threat of being stung by a bee during their beekeeping activities. At first the idea of being stung seems like it would put you off beekeeping forever. However bee stings, although uncomfortable, are not disastrous. This is especially true when we have followed measures to protect ourselves when working with our hives. I promise you, a bee sting does not have to signify the end of your beekeeping hobby. I will show you how to reduce the likelihood of bee stings and effective ways of treating your stings.

Firstly, when beekeepers are properly suited, they reduce their chances of getting stung by bees by a significant proportion. I know of some beekeepers who will wear a full layer of clothing and then proceed to put on their beekeeping suit on top of their clothes. The layers of clothing provide much needed protection, reducing the amount of contact between your body and the bees. As I had mentioned earlier, you can purchase individual pieces of protective gear such as the veil, gloves, or suit or depending on your preference, or you can wear the full protective kit. I would advise beekeepers to refrain from wearing perfume or clothing washed in scented detergents because this odor can be easily picked up by the bees and at times, it could make them excited or agitated.

Our demeanor must also be adjusted when we approach our bees. Animals are great at sensing how we feel and usually they will mirror our own behavior. If we are loud and excitable around our bees, they will also exhibit the same energetic behavior. However if we block out all of the mental noise in our head and calm our mind, we will find that the bees will be more relaxed around us and at times they will allow us to finish our work in peace. Beekeepers must also make slow and deliberate movements, avoiding moving around too many objects at the same time. When approaching the hive, beekeepers must advance from the back or the side of the hive and stay clear of the flight path (at the front of the beehive). I would also recommend that hive inspections be scheduled on sunny days when the worker bees are out working and the hive is less populated.

Finally, all beekeepers should carry a smoker with them when they are on missions to the hives. A smoker is useful because it calms the bees and makes them less irritable when human hands reach into their home. A few puffs of smoke can go a long way. This is because as soon as the smoke enters the hive or the surroundings, it masks the alarm pheromone present in bees which usually triggers their natural defensive mode. The smoker can make cleaning and managing a hive a lot less dangerous because we do not have to deal with hostile bees. Do not be shy to spray smoke every few minutes when you can sense that the bees are interfering with your ability to work quickly and effectively.

If you have been stung by a bee, don't panic. In most cases, your bee sting can be managed and treated without having to visit the local emergency room. The first step to take once you feel a sharp sting is to move away from the hive and go to a place where you can assess your sting safely. You can now remove the bee stinger by scraping it off with a fingernail, credit card, or any accessory that can lift the stinger out of the skin. Ensure that you are scraping it off and not pulling it out. When we pull out a stinger, it creates pressure on the venom sac and thus releases more venom. This will create more pain that we are trying to avoid by all costs.

Once the stinger has been removed, rinse the affected area and clean it with anti-bacterial soap and warm water. You can also apply an ice pack if there's some swelling or discomfort that you're experiencing (otherwise antihistamines will also effectively treat swelling or itching). Monitor the affected area for another 20 minutes and assess whether the pain or swelling is slowly increasing or slowly decreasing. If you are not certain whether or not you are allergic to bee stings, keep an EpiPen emergency sting kit in order to reduce any side effects. However, if you were stung in or on the mouth, eyes, ears, or nose, or if you are experiencing trouble breathing, you should immediately call the hospital emergency line or drive to the emergency room.

Protecting Your Hive

A beekeeper is a bee's best friend because we have the foresight to see danger before it approaches the hive. Our alertness can save our bees from being exposed to deadly diseases and other pests. It is always useful when the beekeeper knows what to look for because it makes protecting our bees a lot easier. Generally speaking, inspecting the hive every so often will allow the beekeeper to identify any new and sudden changes which do not look normal. Always trust what your gut feeling is communicated to you about your bees. If it does not look, sound, or feel normal, it is probably worth further investigation.

Pests and Diseases

Keeping hives healthy is a beekeeper's number one priority. This is not always an easy feat though, because as colonies of bees get transported throughout the country, they carry with them infectious diseases which can cause fatal risks. The more knowledge and awareness that we have about these pests and diseases, the greater our ability to save our bees will be.

The worst enemy to a bee would have to be a varroa mite. These parasites attach themselves on the body of a bee and feed on its blood, effectively weakening the adult bee's strength.

The reason why it can be difficult to treat a hive infested with varroa mites is due to the fact that they reproduce in a honey bee colony by hiding inside a honeycomb cell, and emerging with the colony's offspring.

Tracheal mites, on the other hand, are parasites found inside the tracheal tubes of a honey bee, making it difficult for the bee to breathe. These mites are very difficult to identify because they are microscopic in size and they do not appear on the external body of a bee. However, most of the fumigants used to treat varroa mites are also effective in treating tracheal mites.

Pests can also pose a significant danger to the health of our hives. The small hive beetle (SHB) which originates from sub-Saharan Africa has made its way to the shores of America most likely through bee packages or beekeepers who have migrated to the United States. This little beetle can place a lot of stress on a bee colony, especially those that are already weakened by other parasites and conditions. Usually when the beetles start to multiply in a hive, it becomes an overwhelming experience even for the strongest of bees. The beetles attack and destroy unprotected honeycombs in a very short amount of time, removing the hive's source of food.

A wax moth is another pest which infuriates our bees because they will attack the stored honeycombs. I find that similarly to small hive beetles, the wax moths will only cause catastrophic damage to the hive when the strength of the hive has already been compromised by other factors such as mites or diseases. Therefore, the best defense against wax moths is to ensure that the hive is taken care of and always in good health. We can protect our honeycombs from wax moths by fumigating them, or alternatively we can store our honeycombs in a dry area with plenty of sunlight exposure and cool air.

Honey bees can also suffer from diseases. One of the most contagious diseases that can threaten the survival of bees is American Foulbrood, which is bacteria that causes larvae to die once the cells have been capped. This bacteria is easily spread to other cells when worker bees remove the dead larvae. Soon the colony of bees will grow weaker and the new generation will die out. At this point, the colony is unable to defend itself against foreign intruders. Unfortunately, once a bee hive has contracted American Foulbrood, it cannot be saved; instead it needs to be burned in order not to contaminate the other hives. European Foulbrood is a similar bacterial disease, however it can be treated when detected early. Beekeepers can purchase a prescription of antibiotics and treat their hive until the bees have restored their health.

Honey bees also face many viruses, one of which is the sacbrood virus. This virus attacks the larvae and it can be mildly contagious within the beehive. Most of the time, beekeepers will try to control the virus by replacing the queen. This is because a break in the brood cycle will provide the worker bees with enough time to remove all of the contaminated larvae. This is the only known strategy for treating the sacbrood virus because as of now, there has not been a medical treatment found for curing this virus. In fact, there are no known medical treatments for curing viruses. Therefore, our ability to successfully control mite and pest population in the hive will significantly reduce the presence of viruses.

Pesticide Poisoning

Pesticides are chemicals used to manage pests and disease carriers which may harm plants and crops. Usually gardeners or farmers will spray their plants with pesticides to repel animals, insects, and fungi or bacteria which may grow on the plants. The only problem with the use of pesticides is that our precious bees draw out nectar and pollen from plants as their source of food. When a bee consumes nectar or pollen which has been compromised with pesticides, they can be poisoned and die. Sometimes bees are exposed to pesticides through the water that they drink when the water source has become contaminated. Pesticide poisoning is prevalent especially in high agricultural zones where farming is the main industry. The honey bees will be drawn to the luscious fruit and vegetables, unaware that the crops have been treated with chemicals.

Some of the symptoms of pesticide poisoning include finding dead bees on the outside of the hive, having bees die in numbers within days of each other, observing adult bees moving slowly or showing signs of paralysis, or the unfortunate event of a queen dying.

When we see these signs, we must go into full management mode and do our best to try to prevent any further hive deaths. One of the first things we can do is to move our hive into a safe area away from any plants that we suspect might be contaminated. We can also remove stored pollen or honey that may have been contaminated and temporarily choose to feed our bees sugar syrup up until we have reestablished order within the hive. Lastly, we can also replace the bees which had died with new and healthy bees from another hive. In the event that our queen has died, we must purchase a new queen immediately or naturally allow the hive to select their new queen.

Once we are confident that our bees are safe, we can now investigate the cause of the pesticide poisoning. Perhaps you have neighbors who are farmers or avid gardeners; you can walk over and ask them questions about how they protect their plants from pests and what substances they use to do so. It would be great, if possible, for beekeepers to collect samples of their infected bees in order to test for any traces of chemicals in a laboratory. This is so that we can be sure that the cause of death was from a pesticide before we blame our neighbors unduly. If we find that our bees were poisoned, we should negotiate with our neighbors on other alternatives to use on plants which are as effective as chemicals but not deadly.

Managing Bee Issues

Sometimes families are dysfunctional, but it doesn't necessarily mean they don't love each other. Instead, it may be seen as an opportunity for the whole family to receive therapy and talk their matters through with an outside party. Our bees are part of huge families and as such, there can be a lot of dysfunction occurring in the hive. As beekeepers, we do not want our bees to have to live in a stressful environment, and so it is our role as the willing and gracious third party to help our bees regain order in their family.

Laying Workers

When a hive survives a period of time without the presence of a queen, it may develop laying workers. Laying workers are worker bees who produce unfertilized eggs which become male drone bees. You can imagine how disastrous this can be for the hive. The presence of more drones than worker bees means that there are less bees doing the work in the colony. Furthermore, the purpose of a drone bee is to mate with the queen and if this is not possible, the drone's work cannot be done (all they can do is consume the food supplies in the hive).

Eventually, the population of the hive will decrease as there are no new workers being raised and some of the drones will have migrated to another hive in search of a queen. The solution would be to immediately find a queen replacement and restore the natural order of the hive. However, this can prove to be difficult because a hive with laying workers will not readily accept a new queen. They perceive their hive to be functioning just fine without a queen and thus they will kill or contend with any bee attempting to assume that position. So what should beekeepers do to correct this dilemma? I would suggest that beekeepers combine a laying worker bee hive with a strong queenright hive which will not allow any bullying from the laying workers. In fact, as soon as the laying workers smell the pheromones of the queen, their reproductive capabilities will be suppressed.

Handling Aggressive Hives

Our bees are usually gentle or mild-tempered and we are able to interact with them without fearing resistance. However, new beekeepers must be warned that sometimes our bees can be quite moody or downright aggressive toward us. We all know that honey bees are naturally defensive when it comes to protecting their hives, however their behavior can be aggravated by other factors such as being provoked or changing weather conditions. Many beekeepers will agree that in the hot summer months, bees tend to be more restless due to the scorching heat, and this can turn a sweet bee into a grumpy one. No one enjoys being around a grumpy bee who is ready to release their stinger at the slightest movement.

One way that we can deal with angry bees is to replace the queen. The queen is influential in establishing the temperament of the other bees. This is because the behavior of the other bees in the colony is impacted by the pheromones released by the queen. An angry queen is always surrounded by angry bees, therefore beekeepers should replace the queen with a more gentle and soft equal. However, sometimes it is not the queen's fault. The aggression from the bees may be due to the type of bee species we have purchased. Whenever we bring bees into our yard, we must be certain of the traits and characteristics of our bees in order to predict their social behavior. If we found our bees in the wild, we are taking a higher risk because we cannot predict the nature of our cross-bred honey bees.

Other causes of the aggressive behavior are associated with hive politics. For instance, there may be a lack of food in the hive, an overcrowded colony, or an infestation of mites. These issues can create a stressful environment for our honey bees, who prefer to be in a peaceful environment. If the issue is food, we can quickly correct it by feeding our bees. If it is a pest problem, then we can fumigate our hive or treat our ailing bees with antibiotics. If the issue is about a lack of room in the hive, we can move some of our bees into a nearby hive and purchase a new queen to mark her territory in the new hive. Moreover, beekeepers should avoid spilling sugar syrup near or on the surface of the hive because it can attract bees from other colonies who have come to rob some food. Intruders can also be managed by reducing the entrance of the hive, protecting our bees from unwelcome guests.

Common Beekeeping Mistakes to Avoid

There is so much to learn about beekeeping that it is impossible for me to condense all of the knowledge into one book. As I mentioned to you, a lot of what I came to know about this hobby was learned on the job through making mistakes and picking up patterns. It is my desire for you to succeed as a beekeeper and avoid making some of the common mistakes which many of us committed when we were starting out. Of course, nothing beats experience, even if it is gained through making silly mishaps. Therefore, do not beat yourself up if you forget to follow cautionary measures. The truth is that even when you fail, you are still learning valuable lessons.

1. Do not assess the health of a bee colony based on traffic alone

It is common for some beekeepers to take shortcuts when making their inspections of the hive's health. Sometimes beekeepers will observe the energetic activity of the bees outside of the hive and deem the hive as healthy. If, for instance, the hive starts growing in number and there are many more bees roaming around the outside of the hive, the beekeeper will assume that the inside of the hive is also flourishing. This is not always the case. In fact, when a problem is given the time to grow and it is only detected once it has reached the outside of the hive, it is usually too late to remedy the situation. I always advise beekeepers to open the hive and thoroughly check to see how the bees are doing and whether the eggs are healthy. Most mites and viruses will grow internally before they are detected externally, therefore we must pay attention to any significant shifts in the behavior of the bees or the quality of the honeycombs.

2. Do not neglect the queen

Beekeepers should always try and locate the queen and assess her health. It is a cause for concern when the beekeeper cannot locate the queen because this may signal that the queen has been dethroned. Usually, queens are marked by beekeepers before they are placed in the hive in order to be recognizable. A queenless hive is in serious danger of becoming a laying worker hive, therefore we must act swiftly. Some of the signs that our queen may be dethroned include the lack of eggs present in comb cells for a number of weeks, or the population of the hive declining. To spot the difference between worker bee eggs and queen eggs, you can look at the position of the egg. Usually, queens will lay their eggs in the middle of the cell, and a worker bee will lay her eggs on the side. Furthermore, a queen will place one egg into one cell, while a laying worker bee can place multiple eggs in one cell.

3. Do not harvest honey too soon

I understand the anticipation that we all have of harvesting our very first batch of honey, however we can jeopardize the process if we harvest our honey too soon. First, we must never forget that we are only permitted to collect surplus honey from our hive, and usually a surplus amount of honey is produced after the first year of our bees living in the hive (after they have survived the first winter). If we collect honey too soon, we could be robbing our bees of food, causing food insecurity in our hive. How much honey we end up collecting eventually should also be monitored depending on our geographical location. For instance, beekeepers who live in areas where they receive plenty of sunshine most of the year should leave at the very least one deep frame filled with honey for the bees. On the other hand, beekeepers who live in areas where it is usually cold and rainy most of the year should leave at the very least three deep frames filled with honey for the bees.

4. Do not cut all queen cells

Some beekeepers may be caught off guard to see large peanut-shaped queen cells popping up in their hive. Perhaps the prevalence of queen cells makes beekeepers believe that there will be too many queens in the hive which might threaten the position of the current queen. Nevertheless, queen cells are not dangerous, poisonous, or a threat to our hive. Sometimes the presence of a queen cell is part of a bigger strategy that the hive is establishing. The bees may be seeking to replace the existing queen soon and therefore they need to start reproducing a replacement. Other times, we may not have noticed it but the colony is already queenless and our bees are reproducing a queen to do emergency damage control. However, if we already have a strong queen and our hive is healthy, the presence of another queen can lead to bee swarming. In this case, you can use the queen cell to build a new hive or to split an already crowded hive.

5. Failing to recognize a nectar dearth

Nectar dearth is a time when there is a scarcity of nectar in the surrounding environment. This is not a pleasant situation for your bees to be in, because without nectar, their food security is threatened. Sometimes the scarcity of nectar is not due to our surroundings but is a result of other bees or wasps robbing our hive of delicious nectar. When our bees are faced with a shortage of food, they will either die or abandon their hive and form a new home in another environment miles away which offers plenty of food supply. Both situations are undesirable for the beekeeper, therefore we need to constantly be vigilant of how much food our bees have available and further put protective measures in place to protect the entrance of the hive from foreign raiders.

Chapter 5:

Harvesting Honey

"The only reason for being a bee is to make honey. And the only reason for making honey is so I can eat it." - Winnie the Pooh

One of the most charming qualities about bees is their ability to make honey. Is there anything as delicious and nutritious as honey? Not only are these little creatures helping our economy thrive, they are also at work producing healthy food for our well-being. Beekeepers are always waiting for the first sign of a honeycomb being formed in the hive. We will do just about anything to ensure that our honey bees are safe, healthy, and stress-free in their new home in order to make sure that the quality of our honey is of the highest caliber. Once we have crossed the first year mark and we have seen our bees survive their first winter, the surplus honey which is produced in the hive is ours for the taking.

Health Benefits of Honey

Honey is a natural ingredient produced by bees and enjoyed by people because of its sweetness, along with the energy and many health benefits that it promotes. Honey may be consumed in its raw form or added as an ingredient in cereals, desserts and juices, or used as part of a skincare regime or health supplement. In the world today, you will find over 300 different types of honey which vary in color, scent, and flavor depending on the plant species the bees frequently visited to extract nectar.

There is also a significant difference in flavor between processed honey and raw honey. Processed honey refers to honey that is manufactured and has undergone a variety of processes in order to produce a sellable product. The concern surrounding processed honey is that it is over-pasteurized and as a result, it loses most of its nutritional value along the journey from the beehive to the supermarket shelf. However, raw honey is not tampered with a lot. When the beekeeper gathers honey from the beehive, it is usually strained in order to remove any wax or other dirt particles. Thereafter, it can be consumed in its raw form. Other beekeepers who may be harvesting honey for commercial purposes may prefer to place their honey under heat so that they can remove potential pathogens. Of course, when we heat our honey, we are also compromising its nutritional value by removing some of the natural vitamins and minerals found in this food source.

For millennia, honey has been used as a medicinal supplement in homeopathic medicine. Some of the ailments that honey has been proven to effectively treat include inflammation, viruses, and wounds, while promoting weight loss and boosting the immune system. The reason why most people prefer using raw honey as part of their health program is because it is 100% organic, meaning that our bodies are less likely to react negatively when we consume honey. It is gentle to ingest or use as a topical ointment for treating infections, rashes, or other illnesses. These are some of the remarkable health benefits that we can look forward to when we begin to add honey as part of our nutritional plan.

One of the reasons why I love to consume a spoon of honey once a week is because honey is a natural antioxidant. Antioxidants are health substances which protect our cells against free radicals in our body. These free radicals make us age a lot sooner than we would like and if they are left unmonitored in our bodies, they could potentially damage our cells, causing heart disease, cancers, tumors, and other diseases. Honey happens to be full of antioxidants such as glucose, organic acids, or phenolic compounds like flavonoids. Who knew that we could prevent some of the diseases that are most prevalent in our later years by incorporating honey in our diet?

Honey also contains hydrogen peroxide, which is an antiseptic and a great weapon for fighting against bacteria and fungus found in our bodies which could lead to illnesses. For instance, when we are suffering from a throat infection caused by bacteria, we can dissolve some honey in water and drink it as a remedy to disinfect our throats and remove the bacteria. However, sometimes the bacteria is found on our skin and is caused by open wounds. Once again, we can apply honey as a topical treatment to kill the germs in the infected area and to also promote tissue regeneration.

We should keep in mind that the honey used for medical treatment is medical grade, which means that it has been sterilized and inspected before it is applied. Therefore, we must consult our physicians or homeopathic doctors before we can use any form of honey as a medical treatment.

Another benefit that we receive with the regular consumption of honey is a healthy digestive system. When our digestive system is sluggish, we lose vital energy in our body and we become more susceptible to other health risks such as weight gain, stomach ulcers, diabetes, or cancers. Honey is an effective prebiotic which strengthens the good bacteria found in our intestines and thus heals our gut. Honey also increases potassium and water intake which are helpful to reduce the severity or duration of diarrhea.

Honey can also be added in cooking recipes, turning any dull dish into a delicious treat. Many people will use honey as a replacement for sugar when making caramel or sweet desserts, reducing the amount of fatty sugars in the dish. We can also make rich sauces and salad dressings by adding a spoonful of honey in vinaigrettes or barbeque sauces. The next time you're cooking outdoors on the grill, you can try a new basting sauce for your meat consisting of honey, apple cider vinegar, mustard sauce, worcestershire sauce, and lemon juice. You can determine the measurements of each ingredient depending on your preferred ratio of sweet and tangy.

What Can We Make With Beeswax?

The benefits of honey expand beyond food or medicinal treatments. We can enjoy the benefits of honey through the creations that we make from it. There is a growing market for organic and locally produced products because of the increasing social dialogues about living a natural lifestyle, recycling our materials, conserving the environment, and reducing the chemical waste produced during the manufacturing process of most of our factory-made goods and accessories. I see a lot of beekeepers capitalizing on this new wave of producing green products and I must say it is heart-warming. Our primary role as beekeepers is to take care of our bees, however we can also earn a living while doing so. Therefore I encourage the use of creativity in making home accessories or gifts from either beeswax or honey, so that we can prove to the greater society that we can protect nature and still make a living from it.

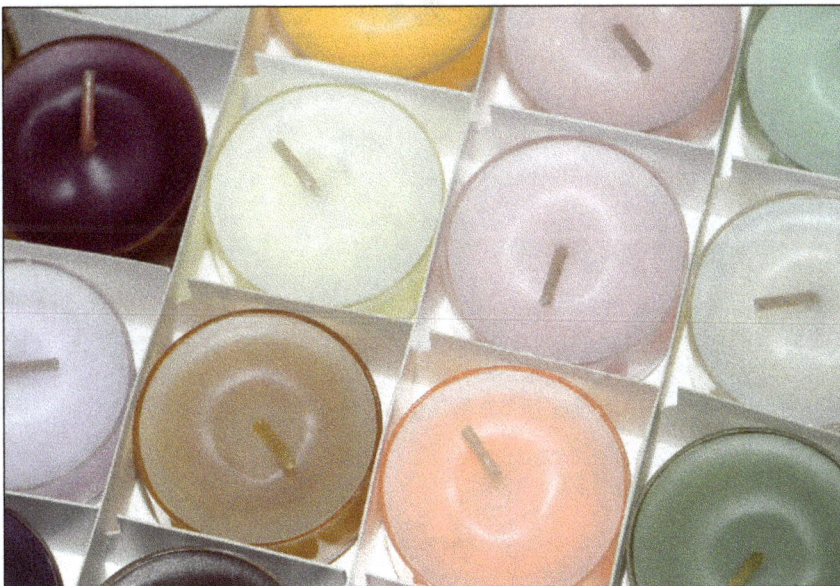

Part of the process of harvesting honey leaves a lot of leftover beeswax. Most bee-keepers will melt this beeswax into various byproducts and sell them online or at a local market. Many people love to purchase products made from beeswax because it is a natural substance that is enriched with a lot of nutrients and it also provides a sweet aroma when placed in our homes. Beeswax also has a long shelf life, it can burn well, and it also boasts protective properties. For instance, many people will purchase candles made out of beeswax that provide a wonderful scent wherever they are lit and are further useful in repelling mosquitoes.

Beeswax is also a natural wood and furniture polish. Rub some of your beeswax on a chair or table that has lost its luster and watch as your furniture piece is revived. This is especially useful when you want to upcycle an older piece of furniture and make it look new again. Your polish mixture can include one part beeswax to three parts of either olive oil or coconut oil (all of these ingredients are 100% organic). Next, heat the ingredients in a bowl over boiling water and stir the mixture until it forms a consistent liquid. Allow the liquid to cool and harden before rubbing some of your polish on a clean cloth and buffing your furniture until it's restored and looking like new.

We can also use beeswax as part of our beauty regimen. I love using natural products for my body because I am certain that they will not cause irritations or adverse side effects that many of our over-the-counter beauty products expose us to. Beeswax is a great ingredient to use in lip balms because it helps our lips lock in moisture and prevents cracked or dry lips. Nothing is worse than having to reapply lip balm several times within the hour because of how useless our lip cream is in providing long-lasting lip protection. Now that you have your own source of beeswax, you can make your own DIY lip balm and will never have to complain about chapped lips ever again.

To make your own balm, combine one part beeswax with one part peppermint oil, coconut oil, and shea oil. Place these ingredients in a bowl and heat over boiling water. You will now have a runny liquid balm which should be placed in a small container and left to cool until it forms a hard balm.

Everything You Need to know About Harvesting Honey

Most beekeepers look forward to the long awaited moment where they can reap some sweet rewards from their honey bees. However, we must remember that there are a lot of factors which can influence the quality and quantity of honey produced in our hive. For starters, we need to understand the species of honey bee that we have purchased. Some honey bees do not produce any honey, while others produce only a little.

Furthermore, we must assess our environment and decide whether there is plenty of nectar within our surroundings for our bees to find and bring back to the hive. The climate in which we live will also impact the quantity of honey that is available for us to collect. Honey bees that live in cold and wet climates will not produce as much honey as those that live in warmer and more tropical regions. The age of our hive plays a significant factor, too. Newer hives will take longer to produce surplus honey. In fact, in the first year of the hive, beekeepers will invest more into the life of the hive than they can expect to reap. When we manage our expectations, we can ensure that we are not collecting honey prematurely or placing our bees under a lot of strain.

When to Harvest?

Before we can consider harvesting honey, we need to make sure that our honey bees are producing a surplus of combs. As a rule of thumb, we should always harvest honey after the winter season has passed in order to avoid robbing our bees of a vital energy source. During the spring and summer months, beekeepers can add more honey supers to the hive so that the bees have enough space to store more honeycombs (this will ensure that the bees are producing more than they can consume). When the beekeeper is ready to harvest, they should first safely remove the honey bees from supers' frames so that we do not kill them in the process. Some of the ways to do this carefully is to use a bee escape board, bee brush, or a leaf blower. Once your bees are removed, inspect the frame and only proceed with harvesting if 90 percent of the comb is capped.

Now it is safe to transport the super frame to a sunny workspace where we will be harvesting the honey about 50 ft from the hive. This workspace can be in a kitchen or garage, however beekeepers must take note that the process of extracting honey can get really sticky and so it is always advised to consider the mess that can be involved. This workspace is also where the frame is left covered with a towel in order to protect it from any bees. Repeat the process of removing frames with as many frames as you wish to extract honey from. Removing frames alone can take 40 minutes to an hour to undertake, so I recommend that beekeepers schedule time to remove frames a day prior to harvesting the honey.

How to Harvest?

Beekeepers can expect their healthy bee colony to produce up to 250 pounds of honey in a single season. However, this also means that during healthy seasons, beekeepers will be hard at work extracting a lot of honey!

An important element that makes this job a little easier is the equipment that we can use during the harvesting process. I know that equipment can be really costly, especially when beekeeping is only a hobby. Therefore, I recommend that new hobbyists rent out the equipment during harvesting season instead of making large purchases during their first few years of beekeeping. Some of the equipment that you can expect to use include an uncapping knife, an uncapping tank, an extractor, as well as a mesh strainer. Below are the detailed steps for harvesting your honey.

Step 1: Pull the frames

The process of harvesting honey begins by removing the frames full of honeycomb from the hive. Make sure that the frames you are pulling are from the honey super and not the brood box. We do not want to disturb any bees during the process of pulling frames, so it's important for us to be aware of what we're moving and to do so with clarity and intention. We must also inspect the frame to see if the combs are healthy and if most of the cells have been capped. Pulling up combs that have not been capped yet means that these combs are still under construction and not fully matured for us to collect.

Step 2: Uncap the honey

Once our frames have been safely removed, we can now begin to uncap them. Remember that our honey is concealed under the yellowish layer of beeswax and in order for us to retrieve it, we must remove this protective layer. There are many methods we can use to uncap our honeycombs. We can use the fork method by utilizing an uncapping fork which looks like a metal brush. To successfully uncap your frames, take the uncapping fork and slide it into each cell and begin to gently remove the wax cover off of the frame. The only disadvantage of using this method is that it can take several hours to individually remove each cell's covering, especially if we are uncapping more than two frames.

Those who would like to speed up the uncapping process can use the roller method. To use this method, you will need to purchase a roller uncapper. When you're ready, roll the roller uncapper over the frame. This will effectively loosen the wax covering on top.

This is a quicker method than using a fork, however it also has its drawbacks. The most noticeable drawback is that as the roller moves across the frame, it can push bits of the wax inside the honey, leaving us with a golden syrup containing wax pieces. It would not be dangerous for us to ingest some wax pieces, although this is not how we enjoy consuming or selling our honey.

Another method we should consider when uncapping our honey is the knife method. This method is the most cost effective because it does not require any special tools, except for a heated and sharp knife. If we are considering using this method, we must plan in advance. Our knife needs a suitable space to run through and slice between the capping and the honey. We can plan for this by allocating enough space between frames in our hive in order to encourage the bees to build the comb out just a little past the edge of every frame. During honey harvesting time, it will give us enough space to slice the wax covering off without dropping any wax pieces into our honey.

Step 3: Place the frames into the extractor

Now that you have successfully removed the wax cappings from your frames, you can now proceed to place each frame into the extractor. As I mentioned previously, there are many options when it comes to purchasing an extractor. The biggest decision you will have to make is whether to use a manual or electric extractor. Most beekeepers start off by using a manual hand-cranked extractor because it is the most affordable option. However, soon the manual process becomes tiresome and time-consuming, especially when the amount of honey being processed is expanding. Nonetheless, hobbyists do not have a lot of honey to harvest and therefore they will not find too much hassle in using a hand extractor.

Regardless of the type of extractor you will use, you must place your uncapped frames into the extractor and either turn the lever round and round or simply turn on a switch and watch as the machine turns the frames on your behalf. As the extractor begins turning, you will see that it removes the honey from the frames and drops it in a pan on the floor of the extractor. This is where your honey will be held and collected until you are able to scoop it out. Once you see that all of the honey has been pulled from the frames, you can transfer it into a food grade bucket for temporary storage and repeat the extraction process on the other frames that are awaiting extraction.

Step 4: Bottle the honey

Now that your honey has been placed in storage buckets, you need to decide what you want to do with it next. You can continue to store your honey indefinitely and keep it until you find a valuable use for it. Honey is a natural preservative and it can be safely stored for up to two years. Remember to check the moisture of your honey before you place it in storage because if it is not brought down to the necessary moisture percentage of 17 to 18 percent, it can ferment and become spoiled. On the other hand, you may also choose to bottle your honey for commercial resale or for consumption at home. I always recommend beekeepers purchase food grade buckets that come with a release valve attached to it which makes pouring into glass bottles a simple and clean procedure.

I prefer placing honey in clear glass containers because everyone enjoys seeing the gorgeous golden color of honey. Furthermore, glass is a quality material which will not compromise the flavor of the honey at all. Unlike plastic, glass will not leach chemicals or odors into the honey. If you intend on bottling the honey for personal use, then I would recommend that you use quart-sized canning jars such as the traditional mason jar. They are reasonably priced, aesthetically pleasing, and do not take up a lot of storage space in our kitchen cupboards or pantries. Each quart jar can hold around 3 pounds of honey. You can also use ½ gallon glass jars which can hold around 12 pounds of honey. I would recommend this option for those who do not intend on moving their honey around too often. These glass jars are reasonably priced and readily available at your local supermarket.

Beekeepers who intend on selling their honey will need glass jars that are more appealing and improve the value of the honey. For instance, you can use an old-fashioned 16-oz muth jar which gives the bottled honey a classic farm style appeal. Individuals who appreciate the antique or traditional culture of beekeeping will find these glass jars complementary to their honey. A great twist on a mason jar would be to use a half pint wide-mouth glass jar. These glass jars are more modern and have a unique shape. I love these jars because they are rarely used by commercial honey manufacturers and so you will not find a lot of them in local supermarkets. The small size of these jars makes them great for gifting friends and family with some of your homemade delicacy.

For bottling, take your glass jar or bottle and place it below the release valve. Slowly open the valve until the glass jar has been filled to a desired amount. Once again, the process of bottling honey takes time, because we want to do it with as much precision as we can to avoid spillage (which can draw bees and create an unwanted crisis). Repeat the bottling process until all of the honey has been transferred into bottles or until you have bottled as much honey as you need. You do not need to refrigerate your honey because it will not expire even when it's left in a dark enclosed room. If you notice that your honey is crystalizing, you can simply place it in a microwave or in a bowl over boiling water and allow it to return to its light golden, syrup consistency.

If you require labels for your jars, make sure that you make them look professional and appropriate to your brand. For instance, if you are a bee-keeper who is very passionate about conserving the environment, you would ensure that your labels are made with 100% organic recycled material and include information about your apiary and what you stand for.

It would be a great investment to have a graphic designer create these labels for a one-off fee, because this label is how your community will come to know about your beekeeping business. Make sure that you check the local or national labelling requirements for honey products so that you can include all of the information that is legally required on labels for the commercial sale of honey. It would also be a great safety measure to check whether your state allows the beekeeper to bottle honey for commercial use at a private residence. Some states will require beekeepers to have a separate honey house where they run all of their commercial honey production.

Tips for Harvesting Honey

Harvesting honey is perhaps one of the most rewarding activities of beekeeping. This is because when we harvest, we are effectively witnessing and tasting the fruits of our labor. All of our time spent caring for our honey bees finally pays off when we spread some of our backyard honey on a warm piece of bread. However, harvesting can be an overwhelming job if we are not mindful of what this activity entails. The process of harvesting honey involves many parts and if we want to complete it with the least amount of trouble, there are some important tips we can keep in mind. Below are ten survival tips for harvesting honey without any hassle or mess involved.

- When approaching the beehive, be calm and pull the frames out gently without drawing too much attention to yourself. Ensure that every move you make is considered and done gently so that your lovely bees do not become irritated.

- Remember that you can only take the surplus honey from the hive. This means that you should only remove frames from the extra honey supers that you placed in the hive. Our bees need to always have an adequate amount of food in order to reduce stress and tension within the bee family.

- Even though I do encourage beekeepers to use a smoker for warding off bees, I do not recommend an excessive use of the smoker (especially during harvesting season) because it can affect the flavor of the honey.

- When removing the wax capping from the frame, do not dispose of the capping because it is a great material to use for making wax products such as candles.

- An alternative to using an extractor would be to use a cheesecloth. To do this, cut out the comb from the frame, crush it, and then follow through by straining the comb through a cheesecloth. The honey should seep through the cloth and the remaining wax can be used for making wax products.

- Give yourself plenty of time. Harvesting honey can take up the whole day and at times it can span over two days. Make sure that you have allocated at least half an hour to remove the honey supers (per hive). Give yourself another hour to set up the workspace where you will do the uncapping and the extractions. You will need a few hours dedicated to extracting and straining the honey, and another two hours for bottling the honey and cleaning up your workspace.

- Instead of harvesting your honey in an open field, I would recommend that you choose an enclosed area with adequate ventilation. I assure you that your nosey bees will smell the honey from miles away and make it their business to find it.

- Harvesting your honey in large batches can save you a lot of time spent setting up your extractor and doing all of the cleaning.

- You can also rely on your bees for clean up. After you have removed what is remaining of your honey from your frames, it will need to be cleaned thoroughly. Thankfully, your bees will enjoy cleaning the frame and storing all of the bits of honey further down in the hive for later use. Simply place your frames back inside the super and on the hive. Within a few days your frames will be spotless. You can also leave other equipment including your extractor or your uncapping forks or rollers near the hive and allow your bees to collect all of the remaining honey.

- Do not be afraid to call for reinforcement. Honey harvesting requires a lot of work, and this may be overwhelming to undertake alone (especially the first few times you harvest your honey). If you have other beekeeper friends, business partners, or neighbors who would be interested in helping you with parts of the process such as bottling or labelling, reach out to them and offer a sweet incentive in exchange for their valuable extra pair of hands!

Chapter 6:
Expanding and Making It Last

The first couple of years as a beekeeper are demanding. I will not try and sugar coat it for you because I believe that if you understand what is expected of you, it will be easier for you to succeed at what many would find to be too much effort. There is a reason why I began the book explaining the requirements of this hobby. I did it so that once you had grasped what a day in the life of a beekeeper would look like, you could decide whether this hobby is for you or not. If you have made it to this final chapter of the book, I can confidently say that you will be one of the many beekeepers who make beekeeping a sustainable and profitable lifestyle.

I have had days along my beekeeping journey where I did not feel like checking up on my little furry friends because I was sour from an unexpected bee sting or confused at the sudden deaths occurring in some of my hives. As the years have gone by, I have become emotionally attached to my bees even though sometimes they see me as an intruder. This hobby turned business has and will continue to be a rollercoaster ride full of highs and lows and a lot of sweet honey to cheer me up. It is my wish that you persevere through the difficult times; when beekeeping becomes expensive, the bees take longer to produce honey, or colony collapse from deadly bee diseases and viruses occurs.

When I get frustrated or overwhelmed at the demands of beekeeping, I remind myself of why I decided to take care of these creatures in the first place. Then, I visit my hives and spend some time watching these beautiful insects as they continue to work. I open the hive and take a moment to appreciate the warm aroma of brood, nectar, and pollen—a smell that only beekeepers know. If I'm really feeling cheeky,

I'll run my finger along a comb and taste the raw honey as it's being prepared by my bees. Honey eaten straight from the hive is nothing like honey eaten from a jar. Eventually, after admiring the bees for a while, I cannot imagine ever quitting this selfless job as a beekeeper. In this final chapter, I will expose you to some valuable knowledge on how to make your beekeeping hobby sustainable for many years to come.

Joining Beekeeping Organizations

I would encourage new beekeepers to make it one of their top priorities to find a local beekeeping organization. This is because we cannot learn about beekeeping all in one book or through attending one course. Our knowledge on how to take care of bees must be refreshed on a regular basis. Being a part of an organization offers us many benefits, such as being able to receive support during our first few years as beekeepers or having access to beekeeping resources and materials which can expand on our existing knowledge. Even as an experienced beekeeper, I always find that I have questions about certain beekeeping processes that I need answers to.

Therefore, I find that any beekeeper, regardless of experience, can gain valuable insights and the much needed support from a local organization. When looking for the nearest beekeeping organization, you can approach your state's Beekeeping Associations. These associations are rich in resources, current news and research, as well as provide access to the legal documents stipulating the requirements and conditions for beekeeping within each state. Some beekeeping organizations will offer training courses and certifications (at a fee) and assist those who would love to develop their beekeeping skills with more advanced training and education.

Beekeeping clubs are similar to organizations, however they are less conservative. A beekeeping association helps beekeepers with registrations, certifications, licensing, marketing, and other professional services within a reasonable membership fee. Beekeeping clubs are groups of like-minded people who congregate to discuss, share tips, and provide support to other club members. There are many questions that can be clarified during club meetings and gatherings, helping the beekeeper feel supported through their journey. There are so many beekeeping clubs around the country and many of them have been in existence for decades. I would recommend that beekeepers search for clubs within their community or state because typically the information shared with club members will be specific for their region and botanical landscape.

The knowledge shared among club members is relevant to the local region and the factors influencing beekeeping within that particular state. For instance, through clubs, beekeepers can learn valuable information about how to successfully take care of bees despite weather conditions, receive information about reputable suppliers, gain exclusive access to locally bred bees, learn about zoning and other imposed legal restrictions, and receive assistance with pest control and managing diseases within the hive. However, I believe that the most significant benefit of joining a club is the personal interactions and connections that are formed with experienced beekeepers. You can read a book and receive wonderful advice, however nothing compares with having real discussions and receiving personal advice straight from the lips of those who have been in the craft for many decades.

Colony Increase

It's inevitable that our bees will grow in number and that our hive will become too small for our bee family. The amazing thing about bees is that they are quite comfortable with us splitting their colony into two or three smaller colonies which we call nucs. This allows the beekeeper the freedom to control the size of each colony and to make the necessary splits whenever it is appropriate. This approach is more affordable than purchasing a new colony of bees when seeking to create a new hive. It will also give us an opportunity to control the genetics of our bees so that the temperament of all the bees is the same.

It would be in the beekeeper's best interest to monitor the growing number of bees within a hive. This is so that the beekeeper can prevent swarming, manage food insecurity within the hive, or decide to expand the number of hives present in their apiary. Sometimes a colony is increased because of a loss of bees during winter months, or the beekeeper's decision to breed bees and sell nucs to other beekeepers.

Our plans for increasing our bee colonies may be noble, however we must take the necessary precautions before we make such a significant decision.

We need to ensure that the parent colony will still have enough food as well as sufficient worker bees to continue the housekeeping within the hive. We should also make sure that there will be enough guard bees who will remain in the parent colony and defend it from unwanted intruders. Lastly, we need to ensure that there are enough nurse bees in the parent colony who will take care of the brood and continue to rear the new generation of bees. Ensuring the health of our parent colony is paramount before we can split it in two.

Expanding our colony of bees is exciting and it is a positive signal of growth in our beekeeping endeavors. For successful splitting, we should always plan in advance before making the actual split. This is so that we can consider all of the factors that will be involved and the implications of our split.

There are also a few considerations we need to be aware of before splitting our colony. First, we need to assess whether or not the parent colony has evidence of diseases such as foulbrood or traces of parasites. If so, we should withhold splitting our colony until we have effectively dealt with the disease issue. The last thing we want to do is transfer diseases from one sick hive to another new and healthy one. Another consideration that we must make is the location of our second colony. It is a well-known fact that when our second colony is positioned closer than 3 miles from the original hive, many of the bees will end up flying back home. Therefore, space becomes a factor when splitting colonies and we must allocate enough of it between our hives.

Moreover, when we move a frame of brood to form part of another colony, we must realize that a singular frame will eventually become three frames of adult bees. This means that we should be conservative with the numbers of brood frames that we transfer because we may soon have overcrowding in our new hive which can lead to swarming. Last, we need to be mindful that when our new colony is too small (or at least smaller in size than our original colony), it can negatively impact the food situation in the hive. Small colonies will have to work harder to find and produce food, and this can place a lot of strain on the food available in the hive. Beekeepers are advised to ensure that splits are made with a reasonably sized colony of bees, large enough to continue the work in the hive and immediately build up their new home.

There are three methods which we should consider that are helpful when seeking to increase our bee colonies. The first one is the most natural way of increasing any bee colony, and that's through catching wild swarms of bees and placing them in a new hive. In order to successfully do this, we must first contact our local beekeeping association and inform them of our desire to catch a swarm of bees for the purpose of expanding our colonies. Of course, when we receive our swarm, we do not know how this colony will behave in our hive. Even though it poses potential risks, we at least save ourselves the hassle of having to split our own colony.

The second method involves making a two-frame nuc, ensuring that the queen is left behind in the parent colony. If our existing hive is healthy and there is plenty of food available in the surrounding fields, removing a small nuc will not cause significant consequences to the production of honey in the parent colony. Within your two-frame nuc, you should add a frame of brood (any more will be too much) and a few extra bees caught from the frames holding food. Do not forget to add foundation or comb if you have enough and relocate the colony further than 3 miles away from the original hive. Add a queen or a queen cell to the new colony once it has settled in.

The third method is to create a modified artificial swarm. This method involves moving the parent colony to a new location and placing a fresh hive (containing a brood box only) in its position. Once the new hive is positioned, remove one or two frames of brood and a frame of food from the parent colony and transfer it to the new hive. Fill the new hive with foundation and give the colony time to become familiar with the new environment. When using this method, do not relocate the queen. Instead, you can place another queen or a queen cell in the new colony.

Making Your Own Split and Nuc

The best time to make your colony splits and nucs is during the late spring months or early summer when there is a strong nectar flow. However, some beekeepers will start making their nucs in the early spring months if they have a fresh queen available and when they are going to use a double-screened board. The purpose of the double-screened board is to help situate the nuc directly on top of the parent hive, in order to allow for the heat transfer from the parent colony to warm up the new colony developing.

To successfully make the split and form a nuc, you will need the regular beekeeping tools and gear, along with the additional boxes that you plan on using to house your new colony. The easiest solution would be to move the nuc into a new hive that you have purchased and prepared for this new family. Alternatively, you can keep your nucleus colony in a 4 or 5-frame nuc box temporarily until you have made the necessary arrangements to place them in a hive. You will also need extra frames to replace those that you've moved from the parent colony into the nuc colony, as well as our trusted friend the smoker.

There are a few guidelines that beekeepers must adhere to before making their splits. First, we must remember that it would be more convenient for us to make the split at midday when the field bees are out working because we want our nucs to be filled with nurse bees and not a bunch of field bees who could potentially resist or attack the new nuc's queen. Remember to always cover the nuc while you're still busy organizing frames to place into them. This will ensure that the bees do not escape from the nuc in your absence and it will also serve to protect the most vulnerable bees in the colony—the brood and the queen.

Now we can begin the process of splitting our colony. Our first move should be to locate the queen so that we know which hive has a queen and which one is queenless. Transport the queenless hive to a new area in order to prevent the nuc bees from leaving their new home and rejoining their parent colony. Thereafter, introduce a queen to the queenless hive by purchasing one or by installing a frame holding a queen cell. You can use one of the methods described above to split your parent colony in the most efficient way for you. Always remember to add foundation or extra honeycomb in order to help the bees rebuild in the quickest time possible.

Increasing Honey Production

One of the major stressors for beekeepers is a hive that does not produce adequate honey. Naturally, honey bees have the ability to produce surplus honey above their required amount for survival. However, the amount of honey produced also depends on other factors such as the area in which the hive is located in the country. When the environment is conducive for working, honey bees can produce between 30 to 60 pounds of honey every year. Bees cannot work without a source of energy and therefore we must consider the nectar flow and available pasture for our bees to gain the necessary nutrients and food source to power up its engine in producing honey. We should ensure that we have great nectar producing plants within a mile radius from our hive and consider moving our apiary 3 to 5 miles away from other apiaries in order for us to not exhaust the same resources.

Besides the availability of food, the health of the queen may also contribute to the quantity of honey produced. The queen is the mother of the hive and everyone living under her roof is sustained by her pheromones. When our queen's pheromones are weak, it means that our colony will also work slower and move without the energy they had when the queen was strong. We may begin to see fewer honeycombs that are built or a reduction in the amount of food collected. At this point, we must plan to remove our queen immediately before any intruders notice and take advantage of our failing hive. As a rule of thumb, beekeepers should replace queens every two years with a young virgin queen that can be reared from within the colony or purchased from a bee supplier. The energy of the young queen will raise the energy within the colony and restore the hardworking and disciplined attitude of our honey bees.

Another reason why our honey production may be low is due to our bees having a slow start in the spring and summer months. I believe that a beekeeper sets the bar and the bees work hard to reach it. Nevertheless, if we set the bar high but do not provide our bees with the necessary resources, our goals will be impossible for them to reach. One of the ways that we can prepare our bees before spring is to feed them sugar syrup approximately six weeks before the nectar flow begins in order to promote brood activity. Ideally, we should have been feeding our bees throughout the winter as well, helping them to preserve their honey and their energy for the spring months ahead. Another precaution that we can take is to insert five or more frames or bars at a time which will give our bees more space to build or hang their combs. Our little friends love to please and typically when they see more storage, it is interpreted as a sign to produce more honey.

Lastly, we need to understand the implications of swarming on honey production. When our bee colony splits and many leave to go and find a more suitable home, we lose a lot of worker bees. Without sufficient worker bees, our hive will struggle to keep up with the food demands. As beekeepers, we need to act immediately. One of the strategies that we can implement is to combine a weaker colony with a strong colony. In this way, our vulnerable colony will have access to all of the resources that the strong colony has. Alternatively, we can take preventative measures and avoid swarming from taking place. Some of the preventative measures include splitting our colony of bees when the hive becomes overcrowded, replacing our queen, moving our hive to an area with nectar flow, and making sure that our queen is healthy and consistently laying eggs.

Bee Removal

Bee removal, as the name suggests, refers to a process of removing bees from a property. This swarm or hive is usually removed by a beekeeper or a pest control company. There are many beekeepers who will respond to calls asking for a hive to be relocated to their apiary or placed safely in the wild. Essentially, bee removals become an opportunity for the beekeeper to increase their bee colonies and to breed more bees. The advantage of adopting a swarm of bees troubling another yard is that they can receive a loving and healthy environment to grow and work in our yards. Therefore, we need to be prepared for the next time we receive a frantic phone call to urgently attend to a terrorizing swarm of bees or beehive.

There are many methods which we can use to remove bees from a property. First, we can do a live bee removal without the use of pesticides. Even though the bees are removed alive, they can face a lot of stress along the process and eventually die. Part of the stress experienced by the bees is caused by the equipment that many bee removal companies will use to remove the bees. For instance, companies will use a heavy duty vacuum cleaner that is not modified to remove bees gently. The force from the vacuum is far too aggressive and many times bees will get pulled up into the chamber of the vacuum and never come out alive. This process is inhumane and can potentially destroy the future of a colony.

The second method is a live bee removal performed in a humane way. The bees are safely caught and released in the wild. The only problem with this method is that the companies will not consider what happens to the colony once they are either dumped in an area without food, dumped in an illegal space, or on someone else's property.

However, there is a third method which is a live bee removal that includes a safe relocation. The person or company performing the removal will safely catch the swarm and relocate them to a beehive box. Even though they are placed in a hive, the company will not ensure that the bees have access to water and food.

The fourth and final method of removing bees is the one that all individuals and companies should seek to follow. This method is the live and humane bee removal which involves the relocation and nurturing of the bees. This method is concerned about the sustainability of the colony. The bees are removed from the property and transported to an apiary, farm, or a beekeeper's home where they will be welcomed with warm hands and a lot of sugar syrup to feed on. The beekeeper will assume responsibility for the survival of the colony and every two weeks, they will perform a hive inspection in order to check the health of the new bees.

Glossary

American foulbrood - A disease found in the hive which is caused by a bacteria killing off the larvae. This bacteria is contagious and can contaminate combs if not detected soon.

Apiary - Hives that are placed in one location for the purpose of beekeeping or bee farming. An apiary can also be referred to as a bee yard.

Apis mellifera - The scientific term used to refer to a European honey bee. When we hear of individuals speaking about beekeeping, we can assume that they are referring to this specific species of bee.

Bearding - This describes the process of bees gathering on the outside of the hive, usually around the entrance. This is done so that the temperature inside the hive can cool down and avoid overheating on hot summer days.

Bee bread - The food substitute given to bees made from a mixture of pollen, honey, and royal jelly. This mixture is prepared by the nurse bees (who are worker bees) for the purpose of feeding the bee larvae, the drone bees, as well as the queen.

Beehive or Hive - A man-made wooden enclosure used to house a colony of bees. One hive is big enough to accommodate one colony of bees.

Beekeeping - The maintenance and handling of honey bees by a beekeeper.

Bee space - The path that is built within a hive which allows for two bees to pass freely at a single time. The space is usually between ¼ to ⅜ inches or 6 to 8 mm.

When the path is smaller in measurement, the bees will fill the gap with propolis. However when the path is larger, the bees will use the space to build another comb.

Beeswax - The substance that honey bees use to build their combs. It is produced by the worker bee when it secretes tiny wax scales from its body and used as a building agent to create hexagonal-shaped comb cells.

Brood - Refers to eggs, larvae, and pupae of all castes within the bee colony.

Brood chamber - The section of the hive dedicated to rearing young bees. It is also commonly referred to as a brood box.

Cappings - Refers to the thin layer of beeswax placed over cells of ripe honey. The layer of beeswax serves to preserve the honey until it is ready to be used.

Castes - The three types of honey bees found in a bee colony: the queen, drone and worker.

Cell - The singular hexagonal unit that is combined with other hexagonal units to form a wax honeycomb.

Colony (of bees) - The entire family of bees living in one beehive, consisting of a queen, drones, workers, and brood in the various stages.

Comb or Honeycomb - A sheet consisting of several hundred hexagonal cells made of beeswax and used to protect brood, pollen, and honey.

Drone - Male bees whose function is to mate with the queen bee and produce fertilized worker bee eggs. These bees do not have any other function in the hive and usually they will die or be chased out of the hive once they have impregnated the queen.

Drone comb - Made up of larger cells than the combs used to rear worker bees.

European foulbrood - A disease similar to the American Foulbrood, however not as deadly. It is caused by a bacteria known as Streptococcus pluton which can be treated with antibiotics.

Extracted honey - Honey that is removed from the comb.

Extractor - A manual or automated machine which helps beekeepers extract honey from the cells of combs.

Field bees - Worker bees that are usually 2 to 3 weeks old that travel outside of the hive looking for pollen, nectar, propolis or water from nearby streams and plants.

Flight path - The direction that bees will fly when entering and leaving their hive. If, for instance, the flight path is facing eastward, the bees will travel east looking for food, and will approach from the east when they are coming back home.

Foundation - The material which beekeepers will provide for their bees in order to give them a good start in constructing their wax combs. Usually this material is made from a thin sheet of beeswax with the impression of comb cells on either side. This sheet is installed into frames by the beekeeper, giving the new bees the much needed guidance on how and where to build their combs.

Frame - Equipment that is made from either wood or plastic which is fitted into hive boxes and designed to hold honeycombs. Typically, frames can be removed from the boxes in order to gather honey or for brood inspection.

Fructose - The primary simple sugar found in honey.

Granulation - The process of sugar crystals forming in honey which may cause it to become solid.

Guard bees - Worker bees that are approximately 3 weeks old which carry a high level of alarm pheromones and venom. These bees will protect the hive against intruders or other incoming bees that come to rob the hive of food.

Hive tool - A metal device used by beekeepers to open hive boxes, pry the frames apart, or scrap propolis or wax off from the hive components.

Honey - Substance produced by honey bees by gathering nectar from flowers and bringing it back to the hive where it is passed from one worker bee to another, each bee adding enzymes to the nectar and absorbing the excess water. A gelish honey is then stored in the comb cells and fanned until most of the moisture is reduced. It is then capped with a thin layer of wax and left to rest.

Honeydew - A sweet liquid substance secreted by aphids on plant leaves and collected by bees as a source of food (especially when the worker bee cannot find any nectar).

Honey harvesting - The process of extracting honey from the combs and preparing it for consumption or commercial resale.

Larva (plural larvae) - Refers to the second phase of a bee's metamorphosis once the egg cell has grown into a white grub-like insect.

Laying worker - A worker bee which lays unfertilized eggs that soon become male drone bees. Laying workers are usually present in a queenless hive.

Mating flight - The flight taken by the queen bee when she mates in the air with several drone bees.

Mead - Fermented honey drink.

Migratory beekeeping - The process of moving beehives from one location to another within a season in order to take advantage of better foraging environments.

Nectar - A liquid substance found in plants which is rich in natural sugars. Nectar is the raw product used in making honey.

Nectar dearth - A period of nectar scarcity.

Nucleus hive (or nuc) - A man-made wooden enclosure consisting of a small colony of bees. This bee colony is usually purchased by a beekeeper in order to establish their first beehive. A nucleus hive typically contains 4 to 5 frames.

Nurse bees - Young worker bees which are responsible for feeding and taking care of developing brood.

Package bees - A large quantity of bees purchased from an apiary contained in a cage with a source of food. Queens are sold separately from drone and worker bees; the two must be integrated with each other over time. Beekeepers purchase packages as a starter pack in establishing their own bee colony in their backyards.

Pheromone - A chemical hormone released from the glands of a bee and used as a form of sending messages to other bees within the colony. Honey bees secrete a variety of pheromones and each one communicates a different message.

Pollen - The protein substance that bees collect from plants and store in the hive as a source of food.

Pollination - The transfer of pollen from the male reproductive cell (anther) in a flower to the female reproductive cell (stigma).

Propolis - A mixture that is produced by worker bees made from tree resins and other plant life. Worker bees use propolis for maintenance inside the hive; filling unwanted gaps, cracks, or holes within the structure of the hive and preventing diseases or parasites from invading their home.

Pupa (plural pupae) - Commonly referred to as "capped brood" because at this stage of the bee's metamorphosis, the larva has been sealed in the comb cell in order to form into an adult bee. The pupa will stay beneath the capping until it has fully formed and is strong enough to chew the capping open and be released.

Queen - A female bee that is fertile and has a fully functioning reproductive system. Queen bees are usually larger and longer than normal worker bees and release a more powerful pheromone than any other bee.

Queen cage - A cage in which a queen is packaged and confined for shipping to a new territory where she will be integrated into a bee colony.

Queen cell - A larger and elongated cell which develops a queen egg.

Raw honey - Honey that is extracted from a beehive which is strained and consumed without adding heat.

Requeen - The process of a beekeeper or a colony replacing the existing queen bee by removing her from the colony and introducing her new replacement.

Robbing - Refers to the act of foreign bees stealing nectar or honey in a hive. This is usually more prevalent during nectar dearth.

Royal Jelly - A substance rich in protein released from the glands of nurse bees and used as a source of food for the queen, bee larvae in their first few days of life, as well as the queen larva until it pupates.

Sacbrood - A viral disease affecting the larvae of honey bees.

Small Hive Beetle (SHB) - A beetle originally from Africa which can cause severe damage to a bee colony and contaminate the combs.

Smoker - Refers to a metal container in which material is burnt to release smoke (and not flames). The cool smoke does not harm the bees; instead it gently controls aggressive bee behavior during feeding or inspections.

Stinger - A structure found on the anatomy of a bee which is used as a weapon for self-defense or offense. Worker bees have a barbed stinger which remains on the body of the animal which is stung. This rip causes the bee to later die.

Super or honey super - Refers to any hive body or smaller box which is used to store surplus combs.

Swarm - The congregation of thousands of honey bees which split from their colony to form a new colony in another location. One of the causes of bee swarming is the overcrowding of bees in a single hive or the lack of sufficient food within the hive.

Varroa mite - The deadliest parasite which can threaten a beehive. This mite will attach itself to a honey bee and feed on its blood until the bee loses its strength and eventually dies.

Virgin queen - A young queen bee that has not yet mated with a drone bee.

Wax moth - The larvae of a moth known as golleria mellonclia which can cause havoc in a hive by emptying the combs or contaminating the developing brood.

Worker bee - Non-reproductive female bees which carry out all of the housekeeping in the hive including searching for food, feeding the colony, producing honey, and protecting the hive from foreign intruders and diseases. Worker bees are the most prevalent caste of bees living within a bee colony.

Conclusion

It only takes a walk in nature to truly appreciate our lives and to see the many ways in which we are blessed. Human beings are given a lot of resources and we have a choice of how to use them in order to survive. However, even though we have been given so much, most of us still find error or fault in how our lives are set up, what we lack, or what was withheld from us. When I reflect on the nature of human beings, my imagination always leads me to wondering how successful we would be if we possessed the mentality of a bee.

A bee is not given much. In fact, the wilderness that was once its home is being destroyed and now more than ever, bees are surviving with the bare minimum.

Yet they are still holding our economy together. Every morning we wake up to fresh fruit, yogurt and cereal—every product an outcome of bee pollination. Honey bees are still powering industries within our economy even though they do not have a place to rest. I am proud that I can say I am one of the courageous few who have heard the cry of bees and responded to it in a swift manner. When I became a beekeeper, I set out to discover a new hobby, not knowing that it would become my life project. I believe many beekeepers start out having the same surface level commitment as me and as time progresses and they fall in love with the personality of their bees, they cannot see themselves ever abandoning the beehive for good.

This book is a collection of all the bits of knowledge, wisdom, and advice that I have received in my five years of beekeeping. There is so much more that I could not possibly fit in this condensed book which you will need to explore and study on your own.

The one impression about beekeeping that I would like for you to remember is that this hobby takes a lifetime of learning to master. The knowledge that you have learned thus far is only the beginning of a continuous journey spent understanding bees, their social behaviors, life inside the hive, and the many ways in which we can all protect the lives of our honey bees. Nowadays, reading material is available online, in libraries, or through beekeeping training courses. Even then, I encourage you to network with other beekeepers within your state so that you can receive the knowledge that comes from pure experience.

You are no Longer a "Newbee"

Your journey as a beekeeper officially starts now. Always remember your intentions for choosing to explore beekeeping because these will be your motivations for continuing the journey when it gets rough. In the beginning, you should expect a lot of administration involved in bringing your first colony of bees home. However, you should not let these legal and technical requirements dampen your spirit. The requirements are put in place to protect you and your bees in order to raise healthy hives that are not a danger to your community. Once you have crossed the initial registration stage, you can now purchase your bees and bring them home!

Raising healthy bees does not come naturally to anyone; it is a skill that is perfected over many years. Therefore, don't be so tough on yourself. It's okay to be nervous when you open your beehive for the first time and it seems as though your bees are angry with you. You will learn their behavior and patterns just as much as they will learn yours. Sometimes you might leave the hive top open or drop a bottle of sugar syrup all over the hive, and these experiences will give you interesting stories to share. After all, beekeeping is a hobby and not an Olympic sport. The mistakes that you make along your journey will provide you with plenty of insight you can use to improve your skills.

Remember that your bees know how to survive. They are a lot stronger than we think. I am always mesmerized by the level of intelligence and endurance that bees exhibit. Their excellence is seen and enjoyed through the many products that are made from the natural material produced inside a beehive. Products made from bees are so versatile that they can be consumed as a source of food or used as decorations, medical treatments, and gifts. After having consumed a spoonful of honey, our bodies are enriched and healing is activated; when we light a wax candle and smell the sweet aroma transcend in the air we feel special. This means that the influence of honey bees is far too magnificent to measure. In one form or another, honey bees continue to remind us of their talent and indispensable work in sustaining our livelihoods.

As beekeepers, it is our mission now to sustain the livelihood of our honey bee. We cannot imagine a world without them, and so together we must continue to strive toward raising healthy bee colonies which multiply in numbers and fulfill their natural function in a peaceful environment. It is up to each beekeeper to make a selfless choice of opening their backyards for these gentle creatures to occupy. The role of the beekeeper, therefore, is significant in the conservation and health of the honey bee. The regular maintenance, feeding, and restoring work that beekeepers perform should not be underestimated; even though it is an act of service which is hardly seen, the difference that each thriving colony of bees makes in our world is evident for all to see.

I sincerely hope that you have found this book meaningful in your pursuit to become a beekeeper. Now it is the time to take a giant leap of faith and prepare for a beekeeping adventure of a lifetime. I wish you the very best in all of your beekeeping endeavors!

References

101 Fun BEE Facts About Bees and Beekeeping. (2017, November 2). Beepods. https://www.
  beepods.com/101-fun-bee-facts-about-bees-and-beekeeping/

A Detailed Look at The Langstroth Beehive - PerfectBee. (2018, December 3). Https://Www.
  Perfectbee.Com. https://www.perfectbee.com/your-beehive/beehives-and-accessories/
  langstroth-beehive-in-detail

About Honeybees - Honeybee Centre. (n.d.). Www.Honeybeecentre.Com. Retrieved July 18,
  2020, from http://www.honeybeecentre.com/learn-about-honeybees#.XxMwxCgzbIU

Abra. (2018, September 3). *What is Bee Removal and the various types?* Bee Best Bee
  Removal. https://www.beebestinc.com/what-is-bee-removal/

Admin. (n.d.-a). *Discovering the Benefits of Beekeeping.* Dummies. Retrieved July 17,
  2020, from https://www.dummies.com/home-garden/hobby-farming/beekeeping/
  discovering-the-benefits-of-beekeeping/

Andreas Schantl. (2019, April 19). *Photo by Andreas Schantl on Unsplash.* Unsplash.Com.
  https://unsplash.com/photos/h7zksspYTkg

Arcuri, L. (2017, October 25). *How, When and Why to Feed Your Honeybees.* The Spruce.
  https://www.thespruce.com/feed-your-bees-3016544

Arcuri, L. (2018, April 23). Everything You Need to Take Care of Honey Bees. *The Spruce.*
  https://www.thespruce.com/supplies-for-beekeeping-beginners-3016775

Arcuri, L. (2019a, June 25). *Which Type of Beehive Is Right for You?* The Spruce. https://
  www.thespruce.com/types-of-beehives-langstroth-top-bar-3016863

Arcuri, L. (2019b, August 15). *How to Keep up With Your Beekeeping Tasks by Season.* The
  Spruce. https://www.thespruce.com/beekeeping-tasks-by-season-3016776

Bee Health. (2019a, August 20). *Beekeeping Protective Gear – Bee Health*. Bee-Health. Extension.Org. https://bee-health.extension.org/beekeeping-protective-gear/

Beekeeper Safety | Bee Health. (n.d.). Beehealth.Bayer.Us. Retrieved July 20, 2020, from https://beehealth.bayer.us/who-can-help/beekeepers/beekeeper-safety

Bees, A. (n.d.). *Beekeeping Terminology*. Amazing Bees | Beekeeper Section. Retrieved July 22, 2020, from https://beekeepers.amazingbees.com.au/beekeeping-terminology.html

Bove, F. (2013, May 9). *Which Bee Is Right For You?* Modern Farmer. https://modern-farmer.com/2013/05/which-bee-is-right-for-you/

Canada Agriculture and Food Museum. (n.d.-b). *Which Foods Depend on Bees? | Bees A Honey of an Idea*. Bees.Techno-Science.Ca. https://bees.techno-science.ca/english/bees/pollination/food-depends-on-bees.php

Canada Agriculture and Food Museum. (2009, October 2). *Pollination | Bees A Honey of an Idea*. Bees.Techno-Science.Ca. https://bees.techno-science.ca/english/bees/the-bee-keeper/pollination.php

Caughey, M. (n.d.). *How to Gather Honey from Beehives*. HGTV. https://www.hgtv.com/outdoors/gardens/animals-and-wildlife/how-to-gather-honey-from-beehives

Caughey, M. (2015, July 14). *9 Tips for an Easier Honey Harvest Keeping Backyard Bees*. Keeping Backyard Bees. https://www.keepingbackyardbees.com/9-tips-for-an-easier-honey-harvest/

Charlotte. (2016, April 17). *Beekeeping 101: Harvesting Honey(b)*. Old Farmer's Almanac. https://www.almanac.com/news/beekeeping/beekeeping-101-collecting-honey

Charlotte. (2019b, June 23). *Hacks for Dealing With Aggressive Bees*. Carolina Honeybees. https://carolinahoneybees.com/aggressive-bees-in-late-summer/

Charlotte. (2020a, February 1). *What to do with Queen Cells*. Carolina Honeybees. https://carolinahoneybees.com/a-queen-bee/

Crystal. (2017a, March 23). *How to Package, Store, and Sell Honey, plus Creative & Inexpensive Honey Labels*. Whole-Fed Homestead. https://wholefedhomestead.com/how-to-package-store-and-sell-honey-plus-creative-inexpensive-honey-labels/

Deeley, A. (n.d.). *Laying Workers*. Beverly Bees. Retrieved July 21, 2020, from https://www.beverlybees.com/laying-workers/

DeJohn, S. (n.d.). *Backyard Beekeeping: A Hobby With Sweet Rewards | Gardener's Supply.* Gardeners Supply. Retrieved July 19, 2020, from https://www.gardeners.com/how-to/ backyard-beekeeping/8529.html

Eufic Admin. (2020b, January 14). *The Health Benefits of Honey and Its Nutritional Value: (EUFIC).* Www.Eufic.Org. https://www.eufic.org/en/healthy-living/article/ the-health-benefits-of-honey-and-its-nutritional-value

Farber, J. (2018). Full Honeycomb. In *Unsplash.* https://unsplash.com/photos/ q5AOruzPg3I

Finding Beekeeping Clubs and Mentors - PerfectBee. (n.d.). Https://Www.Perfectbee. Com. Retrieved July 22, 2020, from https://www.perfectbee.com/learn-about-bees/ about-beekeeping/beekeeping-clubs-and-mentors

Glossary of Beekeeping Terms | Betterbee. (2019). Betterbee.Com. https://www.betterbee. com/Glossary/

Goldman, R. (2015, February 19). *The Top 6 Raw Honey Benefits.* Healthline; Healthline Media. https://www.healthline.com/health/food-nutrition/top-raw-honey-benefits

Grisak, A. (2013, March 26). *Month-by-month Beekeeping.* Hobby Farms. https://www. hobbyfarms.com/month-by-month-beekeeping-2/

Hilary. (2015a, December 7). *10 MISTAKES NEW BEEKEEPERS MAKE.* Beekeeping Like A Girl. https://beekeepinglikeagirl.com/10-mistakes-new-beekeepers-make/

Honey Bee Management. (2015b, October 28). *Ten beekeeping crimes to avoid.* Honey Bee Suite. https://www.honeybeesuite.com/ ten-beekeeping-crimes-you-should-not-commit/

How to make splits & nucs for apiary increases. (2013, June 4). Runamuk Acres Conservation Farm. https://runamukacres.com/how-and-why-to-make-your-own-apiary-splits-nucs/

Jones, C. (2019, March 6). *The Benefits of Joining a Beekeeping Club Keeping Backyard Bees.* Keeping Backyard Bees. https://www.keepingbackyardbees.com/ benefits-of-joining-a-beekeeping-club-zbwz1903zsau/

Kearney, H. (2016, November 5). *How Much Space Does a Beehive Need? Keeping Backyard Bees.* Keeping Backyard Bees. https://www.keepingbackyardbees.com/ how-much-space-does-a-beehive-need/

Kelly Beekeeping. (2017b, October 13). *Thinking About Keeping Bees, part 1 - Costs, Time & Intangibles - Kelley Beekeeping - Blog*. Kelley Beekeeping - Blog. https://www.kelley-bees.com/blog/kelley-beekeeping/thinking-keeping-bees-part-1-costs-time-intangibles/

Layton, J., & Boyer, M. (2010, August 30). *Top 10 Things You Can Make with Honey*. HowStuffWorks. https://recipes.howstuffworks.com/food-facts/5-things-you-can-make-with-honey10.htm

Meggyn Pomerleau. (2020a). Coffee Beans on Brown Wooden Table Photo. In *unsplash.com*. https://unsplash.com/photos/s4uwGNVcll8

Meggyn Pomerleau. (2020b). Person in Yellow Jacket Holding Brown Wooden Box. In *unsplash.com*. https://unsplash.com/photos/ai0mLcXdcBw

Merino, F. (2020). Beekeeper Checking on Bees. In *Pexels*. https://www.pexels.com/photo/beekeeper-checking-on-bees-3578802/

Moeller, F. E. (1980, October). *Managing Colonies for High-Honey Yields | Beesource Beekeeping*. Beesource. https://beesource.com/resources/usda/managing-colonies-for-high-honey-yields/

New Bee University. (n.d.-c). Beekeeping 101: Choosing Beehive Location - NewBee University. *NewBee University*. Retrieved July 19, 2020, from https://www.newbeeuniversity.com/beekeeping-101-choosing-beehive-location/

Nickeson, J. (2019, August 29). *Honey bees*. Honeybeenet.Gsfc.Nasa.Gov. https://honeybeenet.gsfc.nasa.gov/Honeybees.htm

Oct. 24, P. E. |, & 2019. (2019, October 24). *15 Surprising Uses for Beeswax*. PureWow. https://www.purewow.com/home/uses-for-beeswax

Oh My Disney. (2016a, January 17). *7 Winnie the Pooh Quotes to Make Your Day*. Oh My Disney. https://ohmy.disney.com/movies/2016/01/17/7-winnie-the-pooh-quotes-to-make-your-day/

Old Farmer's Almanac. (2016, April 17). *Beekeeping 101: Harvesting Honey*. Old Farmer's Almanac. https://www.almanac.com/news/beekeeping/beekeeping-101-collecting-honey

Orkin admin. (n.d.-d). *What Do Honey Bees Collect: Bee Pollen Collection*. Orkin.Com. https://www.orkin.com/stinging-pests/bees/what-do-honey-bees-collect#:~:text=Honey%20bees%20collect%20pollen%20and

Pasupuleti, V. R., Sammugam, L., Ramesh, N., & Gan, S. H. (2017). Honey, Propolis, and Royal Jelly: A Comprehensive Review of Their Biological Actions and Health Benefits. *Oxidative Medicine and Cellular Longevity*, *2017*, 1–21. https://doi.org/10.1155/2017/1259510

Patterson, R. (n.d.-a). *Colony Increase for Beginners*. Www.Dave-Cushman.Net. Retrieved July 22, 2020, from http://www.dave-cushman.net/bee/increasebeg.html

Patterson, R. (n.d.-b). *Making increase in honey bee colonies*. Www.Dave-Cushman.Net. Retrieved July 22, 2020, from http://www.dave-cushman.net/bee/increase.html

Paul. (n.d.). *What is the best time of the year to start beekeeping?*·Beehour.Com. Retrieved July 19, 2020, from https://beehour.com/what-is-the-best-time-of-the-year-to-start-beekeeping/

Pixabay. (2016b). Assorted Color Pillar Candles. In *Pexels*. https://www.pexels.com/photo/candles-tea-lights-33197/

Ploetz, K. (2013, May 8). *Dear Modern Farmer: How Do I Legally Start an Urban Bee Hive?* Modern Farmer. https://modernfarmer.com/2013/05/dear-modern-farmer-how-do-i-legally-start-an-urban-bee-hive/

Poindexter, J. (2017, July 18). *4 Steps to Harvesting Your Own Delicious Honey in No Time Flat*. MorningChores. https://morningchores.com/harvesting-honey/

Purchasing and installing your bees - PerfectBee. (n.d.). Https://Www.Perfectbee.Com. Retrieved July 19, 2020, from https://www.perfectbee.com/your-beehive/starting-a-beehive/purchasing-and-installing-your-bees

Purdue University. (2016c, July). *Protecting Honey Bees from Pesticides*. Extension.Entm.Purdue.Edu. https://extension.entm.purdue.edu/publications/E-53/E-53.html

Rajendra Biswal. (2020). White and Black Round Textile Photo. In *unsplash.com*. https://unsplash.com/photos/jYT2Np6DIM0

Raposo, J. (2014, August 29). *Everything You Need to Know About Keeping Bees and Producing Your Own Honey*. Www.Seriouseats.Com. https://www.seriouseats.com/2014/08/how-to-raise-bees-honey-beekeeping-introduction.html

Responding to a poisoning event « Bee Aware. (n.d.). Beeaware.Org.Au. Retrieved July 20, 2020, from https://beeaware.org.au/pollination/pollination-and-pesticides/responding-to-a-poisoning-event/

Rusty. (2011, February 18). *How long should I feed a new package of bees?* Honey Bee Suite. https://www.honeybeesuite.com/how-long-should-i-feed-a-new-package-of-bees/

Rusty. (2017c, April 28). *Are you one of the 80% who will quit? - Honey Bee Suite.* Honey Bee Suite. https://www.honeybeesuite.com/beekeepers-will-quit/

samart33. (2017, June 12). *Bee Bonanza.* Askabiologist.Asu.Edu. https://askabiologist.asu.edu/explore/honey-bees

Scott Hogan. (2019). Brown Bees Photo. In *unsplash.com.* https://unsplash.com/photos/OF-YIEWOa7E

Shivakumar, S. (2020). Personal Holding Brown and Yellow textile. In *Unsplash.* https://unsplash.com/photos/Zz-0cw3vAbw

Sustain. (n.d.-e). *Why bees are important | Sustain.* Www.Sustainweb.Org. Retrieved July 17, 2020, from https://www.sustainweb.org/foodfacts/bees_are_important/#:~:text=the%20honey%20bee-

Tentis, D. (2017). Clear Glass Container with Coconut Oil. In *Pexels.* https://www.pexels.com/photo/clear-glass-container-with-coconut-oil-725998/

The Cost of Beekeeping – How Much to Spend? (n.d.). BeeKeepClub. Retrieved July 18, 2020, from https://beekeepclub.com/getting-started-beekeeping/cost-of-beekeeping/

Tracking the Life Cycle of a Honey Bee - dummies. (2016). Dummies. https://www.dummies.com/home-garden/hobby-farming/beekeeping/tracking-the-life-cycle-of-a-honey-bee/

Vaisman, A. : J. (2020, January 8). *LIST: Common Beekeeping Terms You Should Know.* Backyard Beekeeping. https://backyardbeekeeping.iamcountryside.com/beekeeping-101/list-common-beekeeping-terms/

Vaisman, J. (2019, December 14). *5 Tips for Starting Beekeeping.* Backyard Beekeeping. https://backyardbeekeeping.iamcountryside.com/beekeeping-101/5-tips-for-starting-beekeeping/

Zawislak, J. (2015). *Bee Hive Pests & Diseases.* Bee Hive Pests & Diseases. https://www.uaex.edu/farm-ranch/special-programs/beekeeping/hive-pests-diseases.aspx

Zawislak, J. (2019). *About Honey Bees | Types, races, and anatomy of honey bees.* About Honey Bees | Types, Races, and Anatomy of Honey Bees. https://www.uaex.edu/farm-ranch/special-programs/beekeeping/about-honey-bees.aspx

www.ingramcontent.com/pod-product-compliance
Lightning Source LLC
Chambersburg PA
CBHW080602030426
42336CB00019B/3294